# Transport Planning

68, 81-2, 99,
176, 183, 191,
200, 213-4

OTHER TITLES FROM E & FN SPON

**Caring for our Built Heritage**
A Survey of conservation schemes carried out by County Councils in England and Wales
T. Haskell

**City Centre Planning and Public Transport**
B. J. Simpson

**European Directory of Property Developers, Investors and Financiers 1993**
2nd Edition
Building Economics Bureau

**Marketing the City**
**The role of flagship developments in urban regeneration**
H. Smyth

**The Multilingual Dictionary of Real Estate**
L. van Breugel, R. H. Williams and B. Wood

**National Taxation for Property Management and Valuation**
A. MacLeary

**Property Development**
3rd Edition
D. Cadman and L. Austin-Crowe
Edited by R. Topping and M. Avis

**Property Investment and the Capital Markets**
G. R. Brown

**Property Valuation**
**The five methods**
D. Scarrett

**Rebuilding the City**
**Property-led urban regeneration**
Edited by P. Healey, D. Usher, S. Davoudi, S. Tavsanoglu and M. O'Toole

**Transport, the Environment and Sustainable Development**
D. Banister and K. Button

**UK Directory of Property Developers, Investors and Financiers 1993**
7th Edition
Building Economics Bureau

**Urban Regeneration**
**Property investment and development**
Edited by J. Berry, W. Deddis and W. S. McGreal

**Effective Speaking**
**Communicating in speech**
C. Turk

**Effective Writing**
**Improving scientific, technical and business communication**
2nd Edition
C. Turk and J. Kirkman

**Write in Style**
**A guide to good English**
R. Palmer

*Journals*
**Journal of Property Research**
(formerly Land Development Studies)
Editors: B. D. MacGregor, D. Hartzell and M. Miles

**Planning Perspectives**
**An international journal of planning, history and the environment**
Editors: G. E. Cherry and A. R. Sutcliffe

*For more information on these and other titles please contact*:
The Promotion Department, E & FN Spon, 2–6 Boundary Row, London, SE1 8HN. Telephone 071 865 0066.

# Transport Planning

## In the UK, USA and Europe

*David Banister*

**E & FN SPON**

An Imprint of Chapman & Hall

London · Glasgow · New York · Tokyo · Melbourne · Madras

**Published by E & FN Spon, an imprint of Chapman & Hall, 2–6 Boundary Row, London SE1 8HN, UK**

Chapman & Hall, 2–6 Boundary Row, London SE1 8HN, UK

Blackie Academic & Professional, Wester Cleddens Road, Bishopbriggs, Glasgow G64 2NZ, UK

Chapman & Hall Inc., One Penn Plaza, 41st Floor, New York NY 10119, USA

Chapman & Hall Japan, Thomson Publishing Japan, Hirakawacho Nemoto Building, 6F, 1–7–11 Hirakawa-cho, Chiyoda-ku, Tokyo 102, Japan

Chapman & Hall Australia, Thomas Nelson Australia, 102 Dodds Street, South Melbourne, Victoria 3205, Australia

Chapman & Hall India, R. Seshadri, 32 Second Main Road, CIT East, Madras 600 035, India

First edition 1994

© 1994 David Banister

Typeset inTimes $10/12\frac{1}{2}$pt by ROM-Data Corporation, Falmouth, Cornwall
Printed in Great Britain by T J Press Ltd, Padstow, Cornwall

This book was commissioned and edited by Alexandrine Press, Oxford

ISBN 0 419 18930 0

A catalogue record for this book is available from the British Library

**Library of Congress Cataloging-in-Publication data**

Banister, David.
  Transport planning : an international appraisal / David Banister - 1st ed.
      p.     cm.
  Includes index.
  ISBN 0–419–18930–0
  1. Transportation–Planning.     I. Title.
HE151 . B35   1994
338'. 068–dc20
                                            93–33231
                                            CIP

∞ Printed on permanent acid-free text paper, manufactured in accordance with ANSI/NISO Z39. 48–1992 and ANSI/NISO Z39. 48–1984 (Permanence of Paper).

*For Lizzie, Alexandra and Florence*

# Contents

# Preface

*The dominant economic fact of our age is the development not of manufac-
turing but of the transport industries. It is these which are growing most
rapidly in volume and in individual power.*
Alfred Marshall, 1890

One of the most frequent questions raised by my students has been what book
would you recommend for this course. My answer over the last 10 years has
been consistent, namely that I cannot recommend one book. Several books
are useful, but none seems to present a well argued case as to how transport
planning evolved, what are its strengths and weaknesses, what has it achieved,
how does it relate to actual policy decisions, and where is it likely to go in the
future. This book addresses all these issues so that this important gap in the
literature can be filled.

However, the book does not attempt to cover all aspects of transport
planning, but it concentrates on the processes by which analysis links with
policy and the changes which have taken place over the last 30 years. Most
emphasis is placed on passenger transport (public and private) and on land
based modes. Although the context taken and the perspective given is from
Great Britain, extensive coverage is presented of experience in Europe, the
USA and elsewhere to establish whether lessons can be learnt. The importance
of the issues raised by transport planning is not restricted to one country, or
to one policy-making context, or to one analytical approach. Cross cultural
analysis allows very different attitudes and perspectives to be brought together
on common problems, and increasingly solutions are seen as being international, with analysis tools also being transferable between countries and cities.
Attention is also given to the new transport technologies, the environment,
and important policy issues such as market and planning failures. Case study
material is taken from all scales, from the local to the citywide, from regional
to national, and from one country to another country.

The presentation of the book has been structured in three main parts. The first is a *Retrospective Analysis* of the development of transport planning as an important and legitimate area of research and practice, together with its rejection and re-emergence. Experience here is taken from Great Britain and the USA, and direct parallels are drawn with the similar developments in planning analysis. Links are also drawn with evaluation methods in transport and planning, and the instrumental role which they often played in several of the major transport decisions in the 1970s. This part ends with a review of the radical policy alternative presented by the market approach to transport provision, with reductions in public expenditure, regulatory reforms, and the reduction in state ownership of transport enterprises.

The second part of the book is a *Comparative Analysis* of experience in three European countries (Germany, France and the Netherlands) and the USA, and it covers both transport planning and evaluation. In each case, different approaches have been developed in response to national policy issues and the different cultural and analytical traditions. The more recent developments in large-scale integrated transport models and in behavioural analysis are also presented here, as much of the work is international in scope. Again, different countries have been more or less receptive to the adoption of new ideas. It seems that important common issues are likely to lead to a renaissance in transport planning on an international scale. Many of the major links forming part of the new infrastructure are international. Decision-making, financing, construction and operation will all be carried out by transnational agencies.

The third part provides a *Prospective Analysis* of the key issues facing transport planners into the next century. Certain contextual changes relating to demographic, technological and economic factors are all likely to have a significant effect on transport demand, and there is a new imperative to replace existing infrastructure, to build new infrastructure, and to ensure the optimal use of existing infrastructure. All these issues will necessitate different analytical approaches, particularly where new forms of financing are required. The role for transport planning is likely to change as the new political relationships between the state and market are stabilized. Some form of strategic vision is required together with a planning framework within which the market can operate. The role of the transport planner will be to input to this process at a variety of scales – to establish a clearer understanding of the links between longer-term economic growth, shorter-term fluctuations in the monetary economy, and transport demand; to act as a promoter of economic development; to ensure the city regains and maintains its position as an attractive place in which to live and work; and to provide assistance where the market fails, in particular by meeting social needs. As Buckminster Fuller prophetically stated, 'Hope in the future is rooted in the memory of the past, for without memory there is no history and no knowledge.'

# *Acknowledgements*

Many individuals have helped in bringing this book together, and I would like to thank them all. This includes colleagues at University College London, in particular Ian Cullen (now with Investment Property Databank) and Roger Mackett, and others from many European research centres. Alain Bieber and Peter Nijkamp commented on Chapter 5, and Peter Hall has provided an invaluable sounding board for ideas. He was also prepared to read the whole manuscript in draft. Others have helped with the production of the text and diagrams – Lizzie, Paulo, Ann and Jennifer. The research was funded by the UK Economic and Social Research Council as part of their Transport Research Initiative (1987–1991).

*Chapter 1*

# Transport and Travel
# in the Last Thirty Years

## The Era of the Car

The car has revolutionized the way in which we look at travel and communications. Before the advent of mass car ownership in the 1960s, people travelled
short distances by foot or bicycle with the longer journeys being made by bus
or, occasionally, by rail. Life was centred on the locality in which one lived,
with work, schools, shops and all other facilities being available locally. Travel
outside the community was only undertaken for special reasons to visit
relatives or to go on holiday. In 1960, each person travelled on average some
5,600 km. By 1990 that figure had increased to over 12,400 km (table 1.1). In
the 1990s no one thinks twice about using the car to travel to work, to do the
shopping, or to take the children to school. It is used as a form of transport,
as a local bus service, as a goods vehicle and as a statement of status. It is often
the second largest single item of family expenditure after the home and it is
treated with respect and affection. It is generally believed to be the most
desirable form of transport and will normally be used as the preferred form of
transport, no matter how attractive the alternative might be. The user can
always think of a reason why the car is necessary for that particular journey.

Subjectively, these arguments are clear, but objectively the dominant preference for the car is (or should be) less clear. Cars are a major cause of
premature death in all Western countries. In Britain 5,400 people are killed on
the roads each year, with a further 63,000 people being seriously injured (1989),
yet this is a cost that society seems prepared to accept. Transport is a major
user of energy with all forms of transport consuming some 20 per cent of total
energy consumption in Britain. Transport is a major contributor to environmental pollution and is one sector where most of the trends are in the wrong
direction, with increases in emissions of greenhouse gases (some 20 per cent of
carbon dioxide), its contribution to acid rain (45 per cent of nitrogen oxides

and small amounts of sulphur dioxide) and other gases which have effects on morbidity, fertility and mental development (50 per cent of lead and some 85 per cent of carbon monoxide).

The total amount of travel by road (measured in billion passenger kilometres) has increased in the last 30 years by almost three times with only 8 per cent of that figure being explained by population increase. The remainder is due to increases in mobility which relates to the amount of travel in terms of the numbers of journeys and the lengths of those journeys (table 1.1). Whatever the criteria used, the result is the same, namely the relentless growth in the amount of travel. The 1960s had increases of about 130 billion passenger kilometres of road travel, with the 1970s having a lower growth of 90 billion passenger kilometres. But in the 1980s there was an increase of 200 billion passenger kilometres. Across the whole period it has been the growth in car travel which has explained almost all the increase. Travel by bus and coach and the pedal cycle has declined by about 44 per cent, and as a proportion of all distance travelled these three modes now only account for a mere 8 per cent. Rail travel has remained remarkably stable over this period and there has been a rapid growth in air travel.

This increase in mobility reflects the increased motorization and the changes in location of facilities, which have become more dispersed and distances between them have increased. These longer journeys make it less convenient to walk or cycle, or to use the bus. Particularly over the last 10 years, it has become increasingly difficult for planning authorities to refuse planning permission on the basis that access is inadequate. Many authorities, including development corporations, have attracted development through the promise of free parking on sites which are often only accessible by car. There seems to be a basic inconsistency between a strategy for location decisions which encourages greater use of the car and longer journeys, and one that provides local accessibility and is compatible with the need to reduce the use

TABLE 1.1 Increases in travel in Britain, 1960–1990 (billion passenger kilometres).

| Mode | 1960 | | 1970 | | 1980 | | 1990 | |
|---|---|---|---|---|---|---|---|---|
| Bus and coach | 79 | 27% | 60 | 15% | 52 | 11% | 46 | 7% |
| Car, taxi and motorcycle | 157 | 54% | 303 | 75% | 398 | 81% | 597 | 86% |
| Pedal cycle | 12 | 4% | 4 | 1% | 5 | 1% | 5 | 1% |
| Road Total | 248 | 86% | 367 | 91% | 455 | 92% | 648 | 93% |
| Rail | 40 | 14% | 36 | 9% | 35 | 7% | 41 | 6% |
| Air | 0.8 | 0.3% | 2 | 0.5% | 3 | 0.6% | 5 | 0.7% |
| Total | 288 | | 405 | | 493 | | 694 | |
| Average distance per person per year | 5,640 | | 7,500 | | 8,900 | | 12,440 | |

*Source*: Department of Transport (1992*a*).

of resources. Increasingly transport plays a central and crucial role in determining who gains access to what opportunities, and as such forms an important part of every person's quality of life.

The explanation for this phenomenal growth in travel, and in particular the increasing dominance of the car, is complex. Part of the explanation is the decentralization of jobs and other activities from the city centre as a result of the changing industrial base of the country and because of the high land prices in city centres. Population has tended to follow this movement with extensive suburbanization and, more recently, the desire to live out of the city altogether. Cities have become less dense and all of these changes have been facilitated by the car. The second major explanation is the growth in the economy and the real increases in income levels over the last thirty years. There is a strong correlation between Gross Domestic Product and car mileage and a weaker relationship between income levels and car ownership (figure 1.1). Car mileage figures are also modified by the real price of fuel. Since 1960, the real cost of fuel has remained constant, but it did reach peak levels in 1981–83 when it was some 25 per cent more expensive than today. The reductions in the costs of motoring, combined with increases in real incomes during the 1980s have led to this growth in road traffic. From the latest forecasts of traffic growth (Department of Transport, 1989*a*), this growth in traffic is expected to continue over the next thirty years with a more than doubling of traffic levels from their 1988 levels.

Coupled with the spatial dispersion and economic growth arguments has been the changing demographic structure of the population. Britain's population is ageing, and today's elderly are the first generation of retired people that has experienced mass car ownership and so can be expected to continue

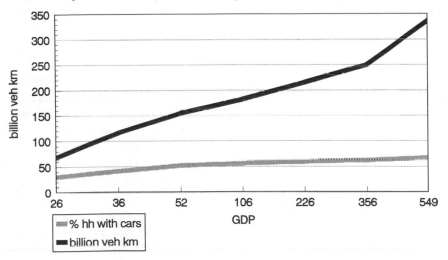

FIGURE 1.1. Growth of Car Ownership and Use against GDP (£bn), 1960–1990.

TABLE 1.2 Summary of the main population trends, 1980–2000.

| Group | Laslett's Ages | Trend | Comment |
|---|---|---|---|
| School leavers and young adults | 1st Age 0–20 | Growth of 10 per cent in the longer term | Decline has already taken place in 1980s and up to 1995 |
| Working age adults | 2nd Age 20–60 | Small increase of 2 per cent | Decline in 20–40 year age group in the 1990s but growth in the 40–60 year age group |
| Age of personal fulfilment | 3rd Age 60–80 | Stable over the next decade | Growth expected in this group in the longer term |
| Full retirement and dependency | 4th Age 80+ | Growth of 25 per cent | Major growth in the next 50 years |

to be car oriented. With the tendency towards earlier retirement, a significant new group is emerging within the population, namely the 'third age of personal fulfilment' (Laslett, 1990). This group (60–80) has ended the complex responsibilities of earning a living and raising a family, they are reasonably affluent, and so have money and time to spend on personal fulfilment. Many of these activities would involve travel and it is here that one major growth area in travel can be expected in the next decade (table 1.2).

It is only when people reach the age of 80 that full retirement and dependency takes place. This group of people, which accounts for about 21 per cent of the elderly, will require special facilities and transport services which can accommodate their particular requirements, for example to be wheel-chair accessible or to have a person to accompany them. This group will not be able to drive and so will require public transport services or taxis or chauffeur driven private cars such as the service currently provided by many voluntary sector organizations. One possible development might be to design a vehicle specifically for the elderly to give them some degree of independence, perhaps similar to the battery operated tricycles which are already available or cars with voice activated functions to ease the physical requirements of driving. Special routes could be provided for low performance vehicles which could be used by the elderly and perhaps children. With the age of consent being reduced (it is now 12 in the Netherlands if there are no complaints), the age at which young people can drive a low performance vehicle may also be reduced, particularly if there is a novice period for newly qualified drivers such as that which operates in Australia. Such a change would allow greater independence for young people and reduce the need for parents to adopt complex escort functions.

In addition to greater motorbility for young people and the growth in the elderly population, the age of retirement is also being reduced, to 55 years or less. This means that apart from the natural growth in the elderly population, the number of people in the retired age group will increase through early retirement, and Laslett's third age would now cover the age range 55–80, not 60–80.

There are also changes in family structure with an increase in divorce rates, single parent families and couples living together without children. These factors together with the fall in fertility rates and the increase in the proportion of elderly in the population have led to smaller household sizes. These demographic trends have an impact on the housing market with an increase in the demand for small housing units which would be located either in the city centres through the subdivision of existing larger properties or in the suburbs in purpose-built units. In each case it is likely that the ratio of car parking spaces to homes will be near unity and that if location is in the suburbs then the number of trips generated and the length of trips will both increase. Saturation levels in car ownership will mean that every able bodied person of driving age will have a car. The actual figure (16–74 years old) will be 71–73 per cent of the total population by 2025, and some 90 per cent of these will have cars – 650 cars per 1000 population. If special cars are produced for the young and the elderly, then this number might increase to nearly 800 cars per 1000 population, about double the present level in most European countries.

## Current Patterns and Trends

The dynamics of change in population mean that the distribution of growth is not uniform. It seems that the link, argued in classical location theory, between home and workplace has been broken as many people now move home for reasons not directly associated with work (Champion *et al.*, 1987). For example, access to a job is not a consideration for retired people; there has also been a growth in the number of people working from home and in the proportion of households with two or more people working in different places; a growth in the numbers of self-employed and the attractiveness of long distance commuting.

The pattern of work journeys needs no longer be fixed by time of day or by day of week. Destinations may vary as individuals visit the head office once a week, regional offices on other days, or are involved with site visits or overseas trips. Time for travelling may also vary. Regular patterns may be established, but not on a daily basis. Commuting patterns have become more complex with cross commuting becoming more important than commuting to city centres. For example, it seems that households now establish a residential base and career needs are met by commuting (Boddy and Thrift, 1990). Within a tight labour market (such as that which has existed in the south of England to 1990), there is often more than one person in each household who is employed, and complex travel patterns emerge as the transport system has to accommodate to this change. With volatile interest rates and little movement in the housing market, it is again the transport system that has to adapt as people cannot move home and so develop long distance commuting patterns. The conditions in the housing market and the increase in the number of economically active people, particularly of women, have led to more travel, longer travel distances and new complexity in travel patterns.

There have also been fundamental changes in occupation and work status with the move away from semi- and unskilled jobs in manufacturing towards information and service based employment. The increased participation of women in the labour force is well known, and this trend is likely to continue with 90 per cent of the expected one million increase in the labour force in Britain being women (1988–2000). Associated with this increased participation of women has been an increase in part-time working and to a lesser extent job sharing. Some of this growth may result from women rejoining the labour force after they have raised their family and from those mature women who have gained new skills through retraining. In Britain the growth in professional, technical and managerial occupations is expected to average 1.7 per cent (1988–2000), more than twice the average for the economy as a whole (0.7 per cent per annum).

As these changes take place travel demand patterns will also respond. Women often have considerably greater constraints on their activities as they have a multi-functional role involving getting the children to school, doing the shopping and carrying out other domestic activities. If employment is added to this list, the number and range of trips generated will also increase. If men take on more of the family responsibilities, more traffic will again be generated. Lifestyles are becoming more complicated for all members of the household not just the adults. Children too are evolving complex social patterns based around the school and home (Grieco *et al.*, 1989).

However, increased participation in the labour force has to be balanced against the shortening of the working week and increased levels of affluence. Economically active people are having to support a growing number of pensioners, as the labour force participation of people over 65 decreases as the level of affluence increases. Again, these trends are likely to generate more travel both nationally and internationally. In the UK 40 per cent take no holidays at all (Government Statistical Service, 1990), over 25 per cent take two or more holidays (1988) and expenditure on leisure now accounts for 17 per cent of all household expenditure. Nearly all those in full time employment have over four weeks of paid leave and 24 per cent have more than five weeks paid leave (1988). The number of overseas holidays taken by UK residents has trebled from 7 million in 1976 to 21 million in 1988, and holidays within the UK numbered 73 million in 1988. The package holiday and changes in air transport have revolutionized the way in which the world is viewed.

Instead of taking only one holiday each year, families and individuals are taking two or three holidays – a main summer holiday overseas, a winter holiday skiing and a third holiday based in their own country. In addition, days are taken off, often at weekends, to give short bridging breaks or long weekends. These trends are likely to continue and increase. It has been estimated (Masser, Sviden and Wegener, 1992) that by 2020 leisure activities may account for as much as 40 per cent of all land transport (in terms of kilometres travelled) and 60 per cent of air travel across all European countries.

The second trend is the increase in affluence and the importance of self-development and achievement. Apart from the activities of the mobile early retired groups, there are many other groups involved in a wide range of activities either of a social or a voluntary nature or of a challenge (for example sporting achievement), or of an environmental nature. In each case time, skill and knowledge is given to this activity and no payment is received. One reason for this self-development has been the growth in real incomes over the last ten years and the increased levels of inherited wealth. As a result of the unprecedented increases in levels of house prices over the last twenty years large amounts of capital are now being spent or passed onto one's children. Alternatively, when individuals trade down in house size or move to cheaper areas, capital is released so that more consumer spending can take place. Similarly, people are borrowing more against the actual or expected rises in house prices. Wealth together with income and available credit have fuelled the increase in consumer spending which has been a feature of the 1980s. Increased rates of growth in car ownership, including a large increase in the provision of company cars, and increased levels of participation in a wider range of activities have all contributed to the growth in the number of trips, the range of destinations and the distances travelled.

In the 1990s these trends have been curtailed through recession which has resulted in falling house prices and home owners with negative equity. Their houses are now worth less than when they purchased them. This in turn makes it difficult to move home and with high levels of unemployment the housing market has stagnated. The impact on the demand for transport has been a slowing down of the expected growth in traffic, and the development of more complex travel patterns. The transport system has to accommodate the new travel patterns generated by a stagnant housing market and people searching for jobs. Once the gearing between house prices and wages is readjusted, the underlying increase in values is likely to continue, but at a lower level. The rapid increases in house prices experienced during the 1980s are not likely to be repeated.

The third change in lifestyles has been the use of the home as an office or workplace. Much has been written on working from home together with the dreams (or nightmare) of the electronic cottage (Moss, 1987; Nilles, 1988; Miles, 1989). It has been estimated that 20 per cent of all urban trips, including about 50 per cent of skilled workers' trips, could be by telecommute, but that only 5–10 per cent of these potential telecommuters would actually make the change (Button, 1991).

The hard evidence of such a change in travel patterns does not seem to be present in Britain or elsewhere. There has been very little change observed in the United States (Hall and Markusan, 1985), although more recent empirical evidence from California does seem to support the argument for modifications to travel patterns (Pendyala *et al.*, 1991). In the California telecommuting project it was found that travel was substantially reduced. On telecommute days, the telecommuters made virtually no commute trips, reduced peak period tripmaking by 60 per cent, vehicle miles travelled by 80 per cent, and freeway use by 40 per cent. As important was the finding that the non-work destinations selected were closer to home, not only on telecommuting days but on commuting days as well. The Californian experiment provides an important indication of the fundamental changes in travel patterns which might be obtained from telecommuting for one or two days per week. This change would reduce commuting and perhaps disproportionately long distance commuting as this is where the advantages of telecommuting are most apparent. However, it is also likely that the long distance commuting journey would be replaced by a series of short distance trips to alternative locations for different purposes. Overall, travel distance may be reduced, but not the number of trips.

These changes in lifestyle identified here will not be felt by all society equally. Some will not be directly impacted at all, whilst others will be, or have been, impacted by each change. It would seem that, as with all social change, it is the affluent that will be impacted first. Those on fixed or low incomes, and those who do not have the knowledge to react to technological change may only benefit in the longer term. The net result of lifestyle changes may be an increased polarization between different groups within society. On the one hand, there will be those affluent individuals with increasing leisure time who will be technologically literate – these information rich people will have increased mobility. On the other hand, there will be poor individuals on low or fixed incomes or unemployed, with no resources or leisure time who will not be technologically literate – these information poor people may possibly have reduced levels of mobility. The distributional impacts of social change may become even more significant in post-industrial society.

## International Comparisons

In a general sense, the pattern of change found in Britain is reflected across the other 11 European Community countries (EC) and those in the European Economic Area (EC countries and EFTA). All countries are moving to the post-industrial era, with increased leisure time, greater affluence and the extensive use of technology. The population in the EC is expected to stabilize at 330 million in the year 2000, but this figure does not, however, take account of international migrations. It is expected that the current low fertility rates will be maintained. The most significant growth will take place in the elderly population, again as a result of the increase in life expectancy and through the tendency to retire earlier. The number of elderly in Western Europe (over 65 years) will rise from 13 per cent (1985) to 20 per cent (2020), and for all Organisation for Economic Cooperation and Development (OECD) countries the number will increase over the same period from 85 million to 147 million. As in Britain, average household size is expected to continue to fall, due in part to reductions in the fertility rate, but also a result of increasing divorce rates and births outside marriage. In the 1970s divorce rates doubled in Belgium and France, and trebled in the Netherlands. By 1986 births outside marriage accounted for nearly half the total births in Denmark and Sweden (Masser, Sviden and Wegener, 1992).

The international movement of labour is still considerable between all European countries and from East to West as well as from South to North. The increased participation of women in the labour force is also apparent in European countries with a similar increase in part-time working. This increase has been particularly marked in the Mediterranean countries (for example, Italy), but the apparent national differences may be due to different attitudes and traditions which are likely to remain.

When patterns of mobility and movement are examined across the EC member countries, similar variation is apparent. Over the ten years 1980–1990 there was an increase of 31 per cent in the numbers of cars and taxis in the twelve EC member countries, with car ownership now averaging 346 per 1000 population. Travel measured in vehicle kilometres and passenger kilometres has also increased by a similar amount with the total amount of passenger travel per head of population increasing to about 8,500 kilometres per annum (table 1.3 and Department of Transport, 1992*a*).

However, the overall pattern has considerable variation and this variation cannot be explained by population, area, density or economic factors such as Gross Domestic Product (GDP) or purchasing power alone. The greatest increase in car ownership over the last ten years has been in the peripheral EC countries with low levels of GDP per capita (Portugal, Greece and Spain). Growth has also been high in Italy, Finland and Japan but mainly from a higher base. Ireland has had the lowest levels of increase, followed by the

Netherlands, Denmark, France, Sweden and Belgium. The other countries, including the United Kingdom, have above average growth (25–35 per cent). Even in the United States the growth in car ownership has been nearly 20 per cent and the assumed saturation level of 650 cars per 1000 has now been reached (table 1.3 and Department of Transport, 1992a). The increase in car ownership in the USA over the decade 1980–1990 has been greater than that of many EC countries, including France, the Netherlands, Ireland and Denmark.

Over the next thirty years there will be a further significant increase in car drivers and in the number of cars (averaging between 60 and 80 per cent) in many European countries, to about 550 cars per 1000 population or similar to the level in the USA in 1979 (Banister, 1992c). This will mean an increase of over 50 million cars in the twelve EC countries (+43 per cent), and much of this growth will take place in households where there is already one car. As in the UK, it is unlikely that the road capacity will increase by anywhere the same amount and so congestion will increase.

Growth has also taken place in car use with Luxembourg, Portugal, Spain and the UK all recording increases of about 50 per cent, well above the EC

TABLE 1.3. International comparisons, 1980–1990.

| Country | Car + Taxi Ownership per 1000 1990 | Growth in Car Owner- ship(%) 1980–90 | Growth in Car Traffic (%) 1980–90 | Average km per Car per Annum 1990 | Percentage by Car 1990 |
|---|---|---|---|---|---|
| Belgium | 393 | 22.4 | 15.0 | 12,200 | 82.0 |
| Denmark | 312 | 15.1 | 30.7 | 17,600 | 79.6 |
| France | 417 | 16.8 | 26.0 | 13,300 | 84.8 |
| Germany | 437 | 32.0 | 35.0 | 11,600 | 85.5 |
| Greece | 159 | 78.7 | – | – | – |
| Ireland | 228 | 4.1 | 30.4 | 24,200 | – |
| Italy | 433 | 39.7 | 36.4 | 10,400 | 79.1 |
| Luxembourg | 483 | 36.8 | 114.3 | 16,400 | – |
| Netherlands | 370 | 14.9 | 30.7 | 14,000 | 84.7 |
| Portugal | 242 | 112.8 | 52.0 | 10,200 | 80.2 |
| Spain | 308 | 52.5 | 50.6 | 6,700 | 74.0 |
| UK | 374 | 35.0 | 56.2 | 16,000 | 88.2 |
| EC 12 | 346 | 31.1 | 37.3 | 13,900 | 82.0 |
| Austria | 384 | 28.4 | 31.8 | 10,700 | 73.8 |
| Finland | 386 | 50.8 | 50.4 | 17,300 | 79.9 |
| Norway | 380 | 25.4 | 54.3 | 15,500 | 87.0 |
| Sweden | 421 | 21.3 | 32.3 | 15,300 | 85.1 |
| Switzerland | 443 | 24.4 | 21.1 | 11,700 | 78.1 |
| Japan | 283 | 38.7 | 44.9 | 9,800 | 54.3 |
| USA | 648 | 18.2 | 39.7 | 15,400 | 98.6 |

*Source*: Department of Transport (1992a).

average of 37.3 per cent. All other EC countries had a growth in car use between 20–40 per cent except Belgium with a figure of 15 per cent (table 1.3). The other countries all recorded substantial increases in car use over the ten-year period. The use made of each car has also increased in many countries, with the UK, Denmark and Sweden all recording more than 10 per cent growth in annual kilometres travelled per vehicle. The only two countries recording a major decline in the use of each car are Portugal and Switzerland. In all developed countries the position of the car is dominant with over three-quarters of passenger kilometres being made by car. The only exception is Japan (54 per cent) where some 37 per cent of travel is by rail. But it is in the UK that one of the largest increases in the use of the car (+49 per cent in passenger transport per head of population), one of the highest increases in distances driven in each car (+13 per cent), and one of the greatest increases in the proportion of total passenger kilometres by car (+5.6 per cent) have been recorded. The EC averages for each of these three indicators are 34 per cent, 1 per cent and 3.5 per cent.

The importance of the transport sector in the EC cannot be underestimated – it accounts for 7 per cent of GDP, 7 per cent of jobs, 40 per cent of public investment, and nearly 30 per cent of energy consumption. These figures only record its direct contribution, and do not take account of transport induced activities nor its role in the overall functioning of a modern society (Group Transport 2000 Plus, 1990).

Growth is also apparent in freight across the EC and other developed countries, but the rate of growth is lower than that of the car. In the EC the total amount of freight tonne kilometres moved has increased by 20.4 per cent, whilst the figure for all nineteen ECMT countries (including the EC) was marginally less. The real growth was in road freight (+34 per cent in the EC) with little change in rail, inland waterways and sea going, but with a substantial decline in pipelines (–14 per cent). From an environmental and energy perspective these trends are disconcerting as the growth has taken place mainly in environmentally damaging road transport which is also energy intensive. The greatest growth rates over the decade in road freight have been found in Belgium, followed by Spain, the UK, Italy and Portugal. The increasing dominance of road freight is expected to continue over the next 20 years with a 74 per cent increase in tonne kilometres carried by roads out of a total increase of 51 per cent (Group Transport 2000 Plus, 1990), and these levels may be underestimates if the Single Internal Market succeeds in eliminating the barriers to movement and in harmonizing taxes between EC Member States.

Modern industry can now locate almost anywhere as it is not dependent on a single source of raw material inputs. Similarly, markets are national and international, not local. One of the results of this location flexibility will be an increase in travel demand. Investment in new infrastructure will improve productivity and reduce transport costs and strengthen the attractiveness of particular locations as there may be considerable economies of scale and scope. The high speed rail network and the Channel Tunnel are two examples of major infrastructure investments which will open up new locations for industrial development. Even in Southern European countries where agriculture is still a relatively large source of employment, the growth in new markets and greater transport accessibility will change the production methods, and increase efficiency and specialization. Companies are becoming transnational in order to exploit local labour cost differentials between individual countries and to compete in world markets (Roos and Altshuler, 1984).

Technological change has reduced the effects of physical distance and allowed further decentralization of lower-order back office functions where cheaper labour can be used, and it is only the front office functions that need to be located in the city centre with a highly skilled and expensive labour force (Goddard, 1989). Many inventory, financial and communication transactions can now be carried out remotely. However, the impacts of technology are not equal across all sectors as they relate to the functions of individual organizations and their links with the high quality and expensive computer infrastructure networks.

High-level knowledge-based activities and skill-intensive tasks may be concentrated in a few core cities and regions, while low-skilled standardized production tasks are carried out in the peripheral areas within individual countries and also within the peripheral areas of Europe as a whole (Masser, Sviden and Wegener, 1992). The level of interaction between organizations both nationally and internationally, and within spatially separated parts of the same organization, is likely to increase by all means of communication, including transport. The greatest unknown at present is the impact of Eastern Europe markets both as opportunities for increased sales and as locations for peripheral factory locations to exploit cheaper labour costs. Growth and affluence may be concentrated in particular countries or in particular areas such as 'the banana' from South East England through Benelux, South West Germany and Switzerland to Northern Italy, or 'the sunbelt' along the northern part of the Mediterranean. The opening up of Eastern Europe will certainly move the centre of gravity of Western Europe to the east.

The uncertainty created by the move to post-industrial organization and the structural changes taking place in industry, together with technological innovation, make it difficult to identify the actual impact of these factors on transport demand. In addition to the economic and technological revolutions, there have been unprecedented changes in the political boundaries with the opening up of Eastern Europe and the Single European Market.

Congestion is probably the most talked about transport issue of the 1990s with 'Gridlock' occurring regularly in city centres, suburban centres and more irregularly in rural areas. Experience in the USA may be informative as car ownership levels are high (650 cars per 1000 people), all forms of gridlock are common and the demographic factors are already being felt. The official view (Federal Highway Administration, 1988) seems to be that the key determinants of future demand are population, age and gender, percentage of driving age population with a licence and personal income. At present nearly 90 per cent of the adult population have driving licences and there are, on average, nearly two vehicles per household. Distance travelled by residents averages at over 29,000 kilometres per annum. These levels are much higher than those in the European Conference of Ministers of Transport (ECMT) countries, and the consequences of the same trends occurring in Europe may be severe as population densities are much higher and the infrastructure is less well developed (table 1.3).

However, not all the evidence is negative. In a most interesting paper, Lave (1990) argues that trend-based analysis is an inappropriate generalization from a highly atypical period of history. As vehicle ownership in the US is reaching saturation and nearly all the driving age population will have vehicle access, the growth rate of vehicle use will decline. He also suggests that the growth rate in vehicle travel will be much lower. While most analysts have been concerned with the consequences of demographic change, they have missed the structural shift in the demographics of car ownership and car use. This shift has led to a disproportionate growth in the vehicle population, but this stage has now ended as the demand for cars is saturated.

However, this optimism from the USA may not be appropriate in Europe, as car ownership has not yet reached the levels found in the USA (table 1.3), and even in the most affluent countries car ownership only reaches between 50 and 70 per cent of the assumed saturation levels of 650 cars per 1000 people. Secondly, there is a link drawn by Lave (1990) between car ownership and car use which he claims is a stable relationship; if there is no further increase in car ownership he argues that there will be no increase in the use of those cars. Evidence from Britain and other European countries suggests that trip lengths have significantly increased, and that the lower mileage recorded by second cars in car-owning households is more than outweighed by the increased mileage recorded by households obtaining their first car. In countries where car ownership is still increasing and where structural changes in the economy are taking place, both the numbers of trips made and the distances travelled will continue to increase, leading to greater congestion.

Within the overall pattern it seems that certain sections of the population may travel more by car. Women and the elderly are two groups which have traditionally driven less than other people. The changes in women's participation rates in the labour force, greater independence and the increase in 'non-standard' households would all suggest that their patterns of travel would become more similar to those of their male counterparts. Similarly, with the growth in life expectancy, health, aspirations and affluence of the elderly, one would expect that they would both keep the car as long as possible and make greater use of it in their extended retirement.

These arguments, at least from a European perspective, would suggest that the complexity of demand for travel will continue to increase. Due to the changes taking place, it is unrealistic to expect that elderly people in the future will have the same travel patterns by mode as a similar elderly group today. There are at least three types of demographic change which would support the argument that trip rates by mode for particular groups will not remain stable in time.

The first factor is that present day expectations and travel patterns will influence aspirations in the future. This cohort effect will be most apparent with the elderly who are the first generation who have experienced mass car ownership and so can be expected to continue to use that mode as long as possible.

The second factor relates to changes in lifestyle, the growth in leisure time and the high value now being placed on the quality of life, and the importance of stage in lifecycle. Lifecycle changes refer not only to the four basic conventional groups (married couples without children, families with young children, families of adults, the retired (Goodwin (1990)), but to the wide range of unconventional groups (for example single parent families). Changes in lifestyle and lifecycle effects have had fundamental impacts on the range of activities that people require, the increasing complexity of travel patterns and the increase in travel distances. Complementary changes have also taken place with the structural changes in the economy and changes in the distribution of industry, commerce and retailing which have tended to follow the decentralization of population.

The final factor has been the increase in levels of affluence and the unprecedented growth in car ownership levels. Some of this affluence has resulted from the growth in Western economies, but the greater part has been the growth in savings and wealth from property value increases. That new wealth is likely to be used by the newly retired elderly or passed on to their next generation. In Britain, it has been estimated by the Household Mortgage Corporation that inherited wealth from the sale of property was £8 billion (1990) and that in the year 2000 the level will be £29 billion (Hamnett *et al.*, 1991). It is unclear what proportion of this money will be invested rather than spent on consumer products such as cars or on other activities such as leisure which involve travel.

In Britain 66 per cent of houses are owner occupied and this contrasts with European levels of owner occupation of 30–40 per cent. There has been a growth in owner occupation across many European countries, including the Netherlands, Sweden, Switzerland and West Germany, during the 1970s and 1980s. In the Netherlands the housing boom was similar to that in Britain but since the mid-1980s house prices have fallen to below 1970 levels (in real terms), and it is only in Britain and Denmark that real increase in prices were apparent to 1990 (Duncan, 1990 [it should be noted that this survey was not comprehensive]). With lower levels of house ownership and lower increases in house prices, the private capital tied up in European housing may be less than that in Britain and so the levels of inherited wealth will be less.

For all these reasons the prediction of travel demand is difficult, but it is clear that demographic changes should form part of the analysis along with historical trends and changes in the economy. The British Department of Transport forecast the general increase in road traffic between 1982 and 1987 as between 9 per cent and 16 per cent, but the actual increase was 22 per cent. Similarly, the GDP forecast over the same period was between 8 per cent and 15 per cent, yet the economy grew by 18 per cent. It was forecast that petrol prices would rise in real terms, but they actually fell (House of Commons, 1990). The present forecasts are for a growth in traffic of between 27 per cent and 47 per cent (1988–2000) and this closely reflects the expected growth in the economy.

The assumption underlying this discussion is that there may be a continued desire to travel, but there may also be a limit to that desire. With the increased levels of congestion on all transport modes, and delays at termini and interchanges, people's appetite for travel may decline, particularly where they have a choice. Quality of life factors become more important with increased affluence and leisure time, and travel may not provide an attractive choice. However, this limit may only be apparent with particular groups in affluent post-industrial western economies. It is likely that any reduction in one group's appetite for travel will be more than compensated for by another group's increased propensity to travel.

It seems that the demand for travel will continue to increase but that the nature of that demand may change as a result of demographic factors. Although the changes in population structure are important, other changes (such as the industrial structure, technological innovation, levels of affluence and leisure time) will also influence demand. The problem here is in unravelling the complexity of issues so that the effects of one group of factors can be isolated. Similarly, there is a range of policy instruments which can be used to influence levels of demand and mediate between the different interests. Two main conclusions arise from this introductory review. Firstly, the overall levels of demand will continue to increase and the private car will accommodate most of that growth. Secondly, the composition of the demand will be significantly different as society becomes less dependent on work related travel and more dependent on leisure travel, and as groups within the population which have traditionally been seen as having low levels of mobility now start to have much higher levels of mobility. The major growth in demand may come from the increased numbers of elderly, the young and women.

The basic policy question then becomes whether and how that increase in demand can be accommodated given the economic, social and environmental costs that will be incurred in developed nations if these mobility trends are allowed to develop unchecked. The alternative must be some form of planning in the allocation of resources and priority to the more efficient modes. The implications for transport operators are considerable if both the demand for public transport is uncertain and the stability of traditional public transport markets is being questioned.

Similarly, there are important decisions for policy-makers in deciding whether to increase the capacity of the road system through new investment or to manage the existing infrastructure through pricing and controls on the use of the car. Action is required, as no action will result in increased congestion and inefficiency in the transport system. Growth in demand takes place continuously, yet growth in the capacity of the transport system is discrete and often takes considerable time for implementation.

The conclusion reached here would strongly suggest that there must be some overall strategic view that links changes in transport demand to changes in demographic factors, changes in the economy, changes in technology and changes in land-use patterns on the one hand, with a concern over the environmental and quality of life factors on the other hand. A single action perspective along one dimension reveals only part of the picture, and to understand and to respond to the whole picture one must investigate the interactions between all relevant sectors to produce a composite strategic view.

## Conclusions

Society is in transition from one based on work to one based on leisure, self-fulfilment and travel. Travel is playing an increasingly important role in our lives with the development of a highly mobile car-oriented lifestyle. Transport planning has also had an important role in understanding the processes of change, in anticipating future growth and in evaluating options. The methods of analysis have evolved from aggregate models applied on a city-wide basis to disaggregate approaches developed to investigate particular problems. In parallel with the development of analysis methods have been the changes in the statutory planning process within which analysis must be placed. Over the last thirty years this planning process has undergone a radical change, with the underlying philosophy switching from the notions of welfare, planning and the availability of transport for all people to one based on the notions of the market, competition and the payment of the full costs of transport by the traveller.

The methods and approaches for the analysis of transport developed in the 1960s and the 1970s were appropriate to large-scale systematic investigation where the responsibilities of central and local government were unambiguous. There was a clear statutory planning framework which was comprehensive in its scope and models could play a central role in the decision process by testing alternatives and evaluating options. Large data sets were available and standard procedures were developed, with transport planning being seen as a technical process. The concern was over the production of proposals supported by comprehensive analysis.

With the reorganization of local government in the 1970s and the widening of expertise, some of the basic assumptions in large-scale analysis methods were questioned. The public wanted transparency in analysis with methods being comprehensible to them and commanding their respect. The notion of the expert and the technocratic nature of the transport planning process was questioned, as were some of the basic assumptions on growth and the absence of what were acknowledged to be key variables (for example employment distribution and land values). This uncertainty coincided with reductions in public funds available for transport investment, and with the restructuring of employment and inner-city decline. Transport planning was no longer primarily concerned with the production of major road schemes, but with evaluation, value for money and the monitoring of change. Coupled with the practical concerns over the usefulness of the comprehensive methods developed in the 1960s and the 1970s, and the changed political priorities, was a fundamental theoretical impasse. It was argued that systems analysis (including transport planning analysis) was functionalism carried out to an extreme and highly complex form (Sayer, 1976). All that had been achieved was the maintenance of the existing social system, and radical alternatives or even the gradual switch to a post-industrial economic base could not be encompassed in such a mechanistic form of analysis.

Transport planning was not prepared for the radical policies of the 1980s with the weakening of the transport and the land-use links, and the emphasis on finance-driven economic policies. The last ten years has seen the demolition of the statutory planning process and the abolition of the Metropolitan County Councils (and the Greater London Council), and the fragmentation of responsibilities in local government. It has resulted in an increased centralization of power with central government having direct control over local budgets. To achieve this magnitude of change there has been a massive programme of legislative reform both in planning and in transport, with the ensuing uncertainty and disruption. The primary aim of the changes has been to reduce the levels of public expenditure in transport, particularly on the revenue support. Where expenditure was available, it should be targeted on capital projects with contributions being sought from the private sector.

Since 1965 there have been three major organizational changes affecting local government (1968, 1974 and 1986). If it is assumed that preparation and adoption take two years each, then 12 out of the last 25 years have been spent preparing for, implementing and adapting to change. If minor changes are also included (1980, 1983 and 1988), then 15 years are accounted for (Tyson, 1991).

The problems facing transport planners are considerable. In the 1960s and the 1970s there was a clear link between policy, planning and analysis, but the 1980s saw a switch to policy-led decisions based more on ideological concerns than analysis. Physical planning as reflected in the land-use and transport planning process is most relevant in a stable environment with public ownership (and a statutory monopoly) of the main transport operators, extensive government intervention, control of prices, and of the levels and quality of service. This approach had its origins in the statutory planning system and the nationalization programme set up after the Second World War as part of Herbert Morrison's plans for the reconstruction of British industry. There were clear welfare objectives in economic policy with the state playing a key role in the reconstruction process. Decision-making was *pluralist* with power being widely distributed between groups without any one group having overall control and the state's role was essentially neutral as it acted as a mediator between the different interest groups.

*Corporatism* developed in the 1970s as a more positive approach to policy-making as well as a response to world recession, high inflation and unemployment triggered off by the rise in oil prices in 1973 and 1979. This period was one of retrenchment where the state formed a series of reciprocal relationships with major organized interests. It seemed at the time to be in the interests of both national and business concerns as the latter were seeking state protection against competition, a defensive mechanism in volatile world markets. Industry operating as monopolies or quasi-monopolies within the domestic sector used the state to plan for growth with minimal competition within either domestic or international markets.

State involvement as partners with industry or as the controlling influence in nationalized enterprises now stood at the centre of decision-making. However, this position was inevitably short lived, as result of both external and internal events. The 1970s were uncertain times in economic terms as trade stagnated, demand fell and unemployment as well as commodity prices rose. The response from industry was to cut back on output, reduce manpower levels and levels of investment. This in turn led to labour disputes and demands for higher wage settlements to match increased prices. At the governmental level taxation was increased, the balance of payments deficit was increased, and the standards of living were at best maintained.

It was against this background that a radical alternative was introduced, namely the move to the *company state* and eventually to the *contract state* where the role of the state is to facilitate the operations of private companies. Initially, the aim was to go towards a market economy based on the well tried neoclassical economic principles. The role of government, both central and local, was to be reduced to that of a facilitator so that the nationalized industries could be returned to the private sector. It was argued that organizations are more efficient in the private sector with the normal commercial pressures that competition brings. Coupled with the privatization programme was a parallel programme of regulatory reform, reductions in trade union powers and the abolition of organizational structures which seemed unnecessarily bureaucratic. The role of the company state is to ensure that the conditions of competition are maintained and that public concerns over such issues as safety and the environment are met. The second stage involves the dismantling of government itself as the role of the state is reduced with the market becoming the arbiter and not the state. This is the contract state where agencies are set up to ensure that competition is fair and that consumer interests are balanced with those of the newly privatized companies. This aspect is particularly important where state monopolies have been replaced by private monopolies.

The basic argument presented in this book is that transport planning must respond to these new imperatives. The approach adopted in the 1960s and the 1970s was appropriate within the stable political environment of that time, but even then in the mid-1970s planning for growth was replaced by substantial concerns over whether the methods were equally appropriate for investigating the transport implications of industrial decline and restructuring. In the 1980s and 1990s society has moved from manufacturing to service employment, from mass production to information technology, to the globalization of the world economy, with major demographic changes in the population and the increases in leisure time. The information revolution is likely to be as significant as the agricultural and industrial revolutions.

The book is divided into two halves, the first presenting a retrospective of the evolution of transport planning. The development of transport planning in Britain is outlined together with its rejection in the early 1970s and its subsequent re-emergence in basically the same form in the late 1970s. This review is complemented by a shorter section on the similar developments in planning analysis. Throughout this period there was an almost parallel evolutionary process. It was during this period that evaluation became increasingly important with growing limitations on public budgets, and the necessity to ensure that value for money was being achieved. The final part of the retrospective presents a view of contemporary transport planning where the market alternative to transport provision is presented, based on the premise that the state should be a facilitator and that intervention should be minimized. Clear conclusions are drawn for transport planning analysis which are intended to act as pointers for the 1990s. These include the issues of infrastructure, congestion, forecasts, regulatory reform, and the environment, all of which suggest that some form of intervention is required.

The second half is a comparison and prospective. The first part explores experience of transport planning in other countries, taking material from the USA, France, Germany and the Netherlands. It also looks at European approaches to evaluation in transport. The purpose of this comparison is to place the British experience in a wider context and to examine the reasons for different approaches to essentially similar transport issues. It examines the processes by which ideas and techniques have developed and explores whether there has been any cross fertilization between countries. The prospective discusses some of the limitations of transport planning and presents some of the new approaches which have been developed in the last five years. Apart from discussing methodological issues, it also comments on the new problems and the role that transport planning can and should have in the 1990s. It sketches out some of the key issues likely to challenge transport planning in the next decade particularly when placed in the context of radical political change, and demographic, industrial and technological changes, and the breaking down of international barriers in Europe. The role for transport planning has changed very substantially over the past 30 years and the prospects for the next decades are summarized in terms of the major priorities for central and local government.

*Chapter 2*

# The Evolution
# of Transport Planning

## Introduction to the Transport Planning Process

Transport planning has evolved over the last 30 years, but with no clear theoretical foundations. Everyone is aware of the problems created by the increased demand for transport and most effort has been directed at finding methods of analysis with a practical, usually quantitative, output. This has meant that analysis has been empirical and positivist in its approach.

Initial developments were concerned with aggregate analysis and the efficiency of overall movement of people and goods. It was in the 1960s that the transport planning process evolved as a systematic method for 'solving' the urban transport problem. The classic deductive approach was adopted with the future state of the system being synthesized from a series of laws, equations and models. The transport planning process was intended to be comprehensive with the collection, analysis and interpretation of relevant data concerned with existing conditions and historical growth. The aim was to establish goals and objectives, to synthesize the 'current patterns of movement' within the city and to forecast future demand patterns either with no changes or with a range of investment options. These alternative packages would be evaluated against the 'do nothing' situation and the goals and objectives set at the beginning of the process.

The structure of the transport planning process followed the systems approach to analysis and marked the move towards an analytical approach rather than decisions being based on intuition and experience. The broad structure of the approach followed that of the Chicago Area Transportation Study (1960), one of the first classic aggregate studies (figure 2.1). This is the basic structure which is still used, albeit with many modifications. Expected vehicle and passenger volumes in the main travel corridors were estimated and sufficient increases in road and public transport capacities proposed to

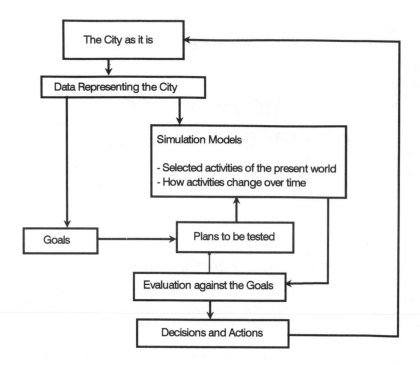

FIGURE 2.1 The Structure of the Chicago Transportation Study. (*Source*: Based on Creighton (1970))

accommodate those expected increases over the following 20 years.

As itemized by Thomson (1974) the basic process can be summarized in eight stages

- *Problem definition*: what is the problem and what are the planning objectives?

- *Diagnosis*: how did the problem originate with views from different perspectives (for example engineering and economic)?

- *Projection*: forecast of what is likely to happen in the future. This is often the most difficult stage.

- *Constraints*: three main types of constraints limit the choice of alternatives (financial, political and environmental).

- *Options*: what are the range of options which can be used to achieve the planning objectives stated in the first stage?

- *Formulation of plans*: a set of different packages covering road and public transport alternatives.

- *Testing of alternatives*: usually through a modelling process to see whether each alternative can achieve the stated objectives and how each compares with other alternatives. Trip generation, trip distribution, modal split and traffic assignment studies.

- *Evaluation*: to assess the value for money usually through some form of cost benefit analysis or financial appraisal.

This structure took several years to evolve, but variations on the basic format have been applied to hundreds of transport studies that have been carried out all over the world (Bruton, 1985; Hutchinson, 1974).

The Transport Planning Model (TPM) formed the central part of the Transport Planning Process and was the testing of alternatives in Thomson's categorization. Conventionally the TPM is divided into four sequential, linked submodels.

*Trip Generation* is the number of trips associated with a zone or unit and consists of trips produced and trips attracted to that zone.

*Trip Distribution* is the allocation of trips between each pair of zones in the study area.

*Modal Split* determines the number of trips by each mode of transport between each pair of zones.

*Trip Assignment* allocates all trips by origin and destination zone to the actual road network. Separate allocations normally take place for each mode.

Other factors such as land-use and population changes are input exogenously to the TPM once it has been calibrated for the existing situation. It is sequential in that the output from one submodel is the input to the next. Information about transport networks, about the location of facilities and about the characteristics of households (for example car ownership and income) are all introduced into the model sequences at the appropriate stages. The output represents the transport system and this is used as the basis against which to evaluate alternative plans. The implicit conceptual foundations of the TPM are that the decisions made by each traveller follow this simple sequence – whether to make a trip, where to go, what mode to use and which route to take. The model does not conform exactly to this decision process as the data are collected for zones and the TPM is carried out for the city as a whole. It is an aggregate model.

## The Development of the Land-Use Transport Study: The 1960s

Up to the 1950s the method by which existing traffic levels was predicted was to extrapolate the growth to the forecast date. A simple growth factor approach was often used giving rise to unconstrained inductive planning. The first real conceptual breakthrough came with the work of Mitchell and Rapkin (1954) at the University of Pennsylvania. They put forward the theory that the demand for transport was a function of land use, and following on from this basic premise it was argued that if land use could be controlled, then the origins and destinations of journeys could be determined – 'urban traffic was a function of land use'. In the mid-1950s large scale land-use studies were carried out on this basis (for example the classic studies in Chicago and Detroit), with changes in land use being taken as the principal exogenous variable. In these large-scale aggregate city wide studies transport planning was seen as a technical study based on what were considered to be rational principles. Plans could be produced on the basis of factual inputs and planners were able to move away from the 'value-strewn area of municipal politics' (Gakenheimer, 1976).

These early studies were developed to assess the changes resulting from new roads, and from new systems of traffic management and control which were being introduced. The early 1960s were marked by exponential growth in car ownership levels, growth in real incomes, cheap energy and a vision of the expansive low-density city. It was also a period of considerable construction with extensive housing programmes and new town development. The optimism of this time was characterized by the view that the reason for congestion was a lack of road space and that this problem could be resolved by more road construction. The early generation plans were heavily road oriented with this naive belief that heavy investment in an upgraded road infrastructure would 'solve' the transport problem.

The first generation of traffic studies was carried out in Britain in the early 1960s. The basic approach has been to model the movement of people and goods in the study area, to identify where congestion is likely, and to predict the use that will be made of the transport network for a 'design' year say 20–25 years ahead, given no change in the supply and given certain specified changes. The alternatives were compared in economic and technical terms with a preferred system covering roads, public transport and parking policies being recommended for the area. These first generation studies involved cooperation between the local authorities, the transport operators and national government. The first and largest of these studies was carried out in London. The London Traffic Survey (1961–62) was intended to only produce road traffic forecasts for 1971 and 1981 and to relate those forecasts to the road networks envisaged for those years (London County Council, 1964). When the Greater London Council came into existence (1965) the narrow objectives of the study were broadened to cover a more comprehensive approach to transport analysis.

These urban traffic studies were complemented by the strategic road plan initiated after the Second World War and brought to fruition through a firm government commitment, between 1953 and 1955, for a basic 1000 miles of motorways. The London-Birmingham (M1) motorway was opened in 1959 with

72 miles of high quality dual carriageway road with grade separated junctions. Its economic benefits had been estimated in a seminal cost benefit study (Coburn, Beesley and Reynolds, 1960), and the political and public impetus for more roads had been created. As Starkie commented (1982) the modified 'tea-room' plan of 1946 had become Conservative Government policy (table 2.3). This period was characterized by the maxim 'predict and provide' with the corollary that cities had to adapt to the car. The shortfall in capacity meant that it was an engineering problem requiring some assistance from the economist on evaluation. The wider implications of road building seemed not to have been discussed and the whole process from problem formulation to prediction, evaluation and implementation was a technical exercise.

Certain events took place in the period 1963–1965 which in retrospect marked a fundamental change in both attitudes to and approaches to the analysis of urban transport problems. The Buchanan Report on *Traffic in Towns* (Ministry of Transport, 1963) made two important contributions. One was the realization that large increases in road capacity tended to exacerbate the problems of traffic congestion and not solve them. Secondly, there was a realization that there were significant environmental disbenefits caused by traffic. Certain ideas were also put forward in the report. The principle of the road hierarchy and the notion of environmental areas formed the core of the Buchanan approach to the design of towns for the motor age. It is ironic that environmental capacity was interpreted by the Germans as instrumental in initiating ideas of traffic calming, but by the British as promoting an expanded programme of urban road building. The German equivalent of *Traffic in Towns* (Hollatz and Tams, 1965) was less concerned about the environment but more concerned about the means to promote traffic flow through the construction of orbital and ring roads, and the necessity for heavy investment in public transport. Hass-Klau (1990) argues that Buchanan was ahead of his time and that it was only with the rise in the environmental movements of the 1970s that the Report's true impact was really felt. However, it could also be said that Buchanan was himself restating ideas, particularly on precincts and their similarity to his environmental areas, which had been put forward to Alker Tripp some 30 years earlier. Nevertheless, it was realized that environmental standards within any urban area automatically determine accessibility and that modification of this relationship would need considerable investment in the physical infrastructure to remould towns and cities for different levels of access (Ministry of Transport, 1963).

The second report received less publicity but was also pioneering in that it examined the possibilities of road pricing in urban areas (Ministry of Transport, 1964). The Smeed Report argued that people should be charged for the congestion they cause. Even though many of the economic arguments were accepted at the time, some are still open to debate now, namely the possibility of very low elasticities of demand, equity concerns and the impact that charging might have on the local economy. The possibility was rejected for technical reasons given the unreliability of metering systems and the costs of implementation. These technical problems have now been overcome, and as with Buchanan's ideas on environmental areas, Smeed's ideas on pricing are now being reconsidered. The problems of public acceptability still remain, but experiences in Singapore, Oslo and Bergen all suggest implementation is possible. No decisions were taken on the road pricing issue and the government eventually produced a report on *Better Use of Town Roads* (Ministry of Transport, 1967) which effectively dismissed both road pricing and supplementary licensing in favour of tighter controls of parking and making better use of the existing road capacity through traffic management schemes.

However, there was now a clear view that traffic studies should not just be concerned with the construction of new roads, but that plans should include the use of the available capacity, public transport options and limitations on the unconstrained use of the car through control measures. This switch to transport studies and the balanced approach to cover all forms of transport was crystallized in the Circular 1/64 issued by the Ministry of Transport (1964). The Ministry would give technical advice and share the costs of these new transport studies. The key issue had become modal split and the relationship between the car and public transport, and the effect of different investment options on the balance between the two.

Even though these studies (for example the West Midlands Transport Study, 1964 [Freeman, Fox, Wilbur Smith & Associates (1968)]; Merseyside Area Land Use Transportation Study, 1966 [Traffic Research Corporation (1969)]) were broader based and included a wider range of land-use alternatives together with public transport networks, their investment proposals were still heavily in favour of roads. Moreover, the assumptions about growth in population and employment on which transport investment decisions were based seemed wildly optimistic. For example, the MALTS study team claimed that the report was based on 'a thorough review of how best to accommodate increases in population and employment over a period of time in which rising productivity and income are expected to lead to a substantial increase in the volume of goods on the road and in the number of cars used.' The MALTS suggested that population would grow between 15 and 28 per cent, and employment by 15 per cent (1966–1991). The main source of new employment would be in the service sector, and would be located in Liverpool city centre (figure 2.2), regardless of land-use policies. Car ownership would rise from 12 cars per 100 people to 40, and 80 per cent of households would have a car.

To accommodate this expected growth, five alternative land-use strategies were examined which combined one of three residential location strategies with either dispersed or centrally located employment. The study team opted for dispersed employment and proposed to accommodate population growth along radial corridors to the east of Liverpool. On the transport side, the MALTS team assumed that people would continue to be allowed to use their

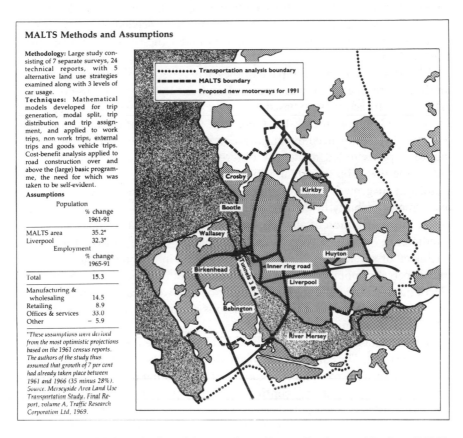

**MALTS Methods and Assumptions**

**Methodology:** Large study consisting of 7 separate surveys, 24 technical reports, with 5 alternative land use strategies examined along with 3 levels of car usage.
**Techniques:** Mathematical models developed for trip generation, modal split, trip distribution and trip assignment, and applied to work trips, non work trips, external trips and goods vehicle trips. Cost-benefit analysis applied to road construction over and above the (large) basic programme, the need for which was taken to be self-evident.

**Assumptions**

| | Population % change 1961-91 | |
|---|---|---|
| MALTS area | 35.2* | |
| Liverpool | 32.3* | |
| | Employment % change 1965-91 | |
| Total | 15.3 | |
| Manufacturing & wholesaling | 14.5 | |
| Retailing | 8.9 | |
| Offices & services | 33.0 | |
| Other | - 5.9 | |

*These assumptions were derived from the most optimistic projections based on the 1961 census reports. The authors of the study thus assumed that growth of 7 per cent had already taken place between 1961 and 1966 (35 minus 28%).*
*Source: Merseyside Area Land Use Transportation Study, Final Report, volume A, Traffic Research Corporation Ltd, 1969.*

Legend in map:
•••••••••• Transportation analysis boundary
∎∎∎∎∎∎∎ MALTS boundary
▬▬▬▬ Proposed new motorways for 1991

FIGURE 2.2 MALTS Methods and Assumptions. (*Source*: Banister and Botham (1985))

cars as and when they chose. Top priority was relief of central area congestion, and this led to the recommendation of a £287 m (1966 prices) roads programme together with a 50 per cent increase in parking provision in the central area.

The road programme consisted of no less than 100 km of new motorway and 90 km of arterial road. The team also recommended building a third Mersey tunnel (the second was then under construction) and thought that a fourth tunnel would be required in the 1990s. Three-quarters of all spending would concentrate on an area within 5 km of the city with an inner ring road forming the top priority.

Public transport improvements were also considered an essential part of the package with the local rail system being upgraded at a cost of £20 m. This involved the construction of an underground loop and link system, connecting the rail termini in Liverpool, together with the electrification of lines entering the city centre. As population was dispersed outwards, commuters were expected to switch from bus to rail; and the £20 m spent on the rail system was expected to reduce the need for spending on roads by some £10 m. By the late 1970s, 4000 park and ride places were assumed to be needed, and 40 trains per hour (in one direction) would be required under the river Mersey (Banister and Botham, 1985).

The MALTS study was typical of the early generation of transport studies in that a careful attempt was made to link transport with land-use planning. The traffic forecasts were based on the best available methodologies of the transport planning process, but there were few signs that the transport planners had been influenced by the land-use planners. Not one of the alternative land-use plans tested made any difference to the need for new transport infrastructure. The reason for this result was that significant city centre employment growth was assumed in all the options. Secondly, to maintain maximum flexibility in the plan, it was argued that the radial roads would allow the distribution of any further growth that might take place above the predicted levels. In retrospect two conclusions can be drawn. Firstly, the inclusion of several peripheral and radial routes in the plan could encourage the dispersal of population and employment, not the concentration assumed. Secondly, it seemed that the ambitious road plans were really a 'bid' for central government resources rather than a serious attempt to promote the best mix of transport and land-use policies.

The decade ended with a further shift in policy emphasis with the 1968 Transport Act. The concern here was with no single best solution but with the allocation of resources and the best use of the existing infrastructure. Traffic management and parking controls were the two main measures used to increase flows and maintain speeds. Every authority over 50,000 population, and all traffic and parking authorities in the provincial conurbations were required to produce short-term Traffic and Transport Plans to cover the period up to the mid-1970s (Circular 1/68 Traffic and Transport Plans and table 2.1). Progress was slow and by early 1972 when they were superseded, only 50–60 out of about 140–150 authorities had submitted plans. It was the plans for urban motorways and other large scale-road schemes which were capturing both attention and the headlines (Starkie, 1973).

## Rejection of the 1960s Approach

The optimism of the early transport studies was overturned in the early 1970s with a movement against transport planning and large-scale road building programmes. The process was no longer just seen as a technical exercise for the expert and his advice could no longer be accepted automatically. Other considerations apart from the technical feasibility and the cost were now considered important. There was a concern over environmental and social issues, for those without access to a car and the assumption that cheap fuel would always be available. This issue was brought into sharp relief in the

TABLE 2.1. Urban transport planning, 1968.

*The Plans* – need for measures designed specifically for

(a)  The relief of congestion
(b)  The help of public transport
(c)  Building road safety measures into highway and traffic plans
(d)  The protection of the environment by traffic management

Urban authorities to prepare traffic and transport plans 'showing how they intend to relate their traffic and parking policies to their available road capacities and to their immediate and longer term policy objectives.'

These plans to be short term (to mid-1970s) and to take account of how other investment programmes (for example urban renewal, highway programmes, public transport, and parking) affect future traffic patterns.

*The Scope of the Plans* – broadly should state

(a)  The present and future situation (traffic, public transport etc) as far as possible in quantitative terms
(b)  The local authority's transport objectives against the town planning background
(c)  The criteria, guidelines and rules of thumb (however crude) which the local authority has applied in arriving at its plan
(d)  Alternative strategies considered
(e)  The chosen plan and its cost

*Source:* Ministry of Transport (1968).

1973–74 War in the Middle East, the doubling of petrol prices, the rises in inflation, world wide recession and increases in unemployment. Economic growth had declined and other quality of life issues were becoming important. The Economist Intelligence Unit had published one of the first critiques of transport studies in 1964. The EIU (1964) argued that forecasts had been based on the unconstrained use of the car in cities, but if new roads were not built to continue this level of car use then conditions would change and some of the forecast growth in traffic would not take place. The key point was that the level of traffic demand is dependent upon the levels of congestion and by implication on the costs of travel.

There was a period of constructive re-examination prior to a vitriolic attack on the use of the systems approach to the analysis of transport problems. It is interesting to note that a similar attack on the use of systems approaches for planning analysis had taken place both in the USA (Lee, 1973) and in Britain (Batty, 1976). The re-examination criticized the narrow conceptual base of the methods, the inflexibility in the basic structure of the process, the lack of any real theoretical development, and the fact that the models did not seem to be responsive to the changing demands of the politicians. Lee's (1973) requiem for large-scale models suggested that the goals had never been achieved and often they had not been stated at the beginning of the study. For example, public transport had been promoted as a reaction against the impacts of highways and not as a real alternative in their own right. Lee also argued that for each objective offered as a reason for building a model, he could suggest a better way of achieving that objective (for example more information at less cost) or propose a better objective. It was a plea for the use of simple methods, for starting with a problem that needs solving not a methodology that needs applying, and for a balance between theory, objectivity and intuition.

In many cases it was accepted that the criticisms and proposals were valid, but it was also argued that the scale of the attack was unjustified. Kain (1978) stated that most criticism had taken place against some normative ideal. Criticism should be placed within the context of alternative problem solving frameworks. Most of Lee's comments were directed at the situation in the USA where the scale of the transport studies was considerably larger than those in Britain, where order, model consistency and comprehensiveness in approach were more apparent.

However, apart from the changes in the economic climate and the move away from road building towards management solutions, there were concerns over the assumptions made in the studies, the scale of the approach and the output. The expectations of growth in population, employment, incomes and car ownership were all very optimistic, and the studies assumed that most growth would take place in the city centre, not at peripheral and greenfield sites. If the MALTS study is taken as an example it can be seen that population within the core was declining even before the study was completed. Between 1966 and 1971 the centre lost 34 per cent of its jobs, a much faster rate of loss than in the rest of the conurbation. Car ownership growth assumptions were also in error as ownership levels reached one car for every five people in 1981, only a half of the forecast figure for 1991 and work trips declined by 10 per cent from 1966 to 1971. If the basic assumptions were proving wrong even while the plan was being prepared, it is hardly surprising that the forecasts also turned out to be wrong.

Excluding the tunnel, the number of trips into the city centre fell by 7.6 per cent between 1978 and 1983 whereas they had been predicted to rise in line with the projected increase in employment. The city centre had not become congested. Traffic speeds in peak hours actually rose during the 1970s, and off street parking in central Liverpool was only 65 per cent occupied in 1981. Similarly, the expected growth in rail travel has not materialized. As elsewhere, the numbers of bus passengers declined, and in 1971 they were only just over half their 1961 levels.

More generally, the Merseyside pattern is found in other towns and cities. Evans and Mackinder (1980) carried out a survey of 31 early British transport studies to compare forecast change with actual change over a ten-year period. The error on average was 12 per cent for population and employment, and over 20 per cent for income and car ownership. They came to the conclusion that if no change had been assumed the results would have been no less accurate! There was some evidence that the models themselves had worked reasonably well but it was the prediction errors in the exogenous variables which had caused overprediction of the increases in demand. Probably more important than these technical reasons was the expectation among the politicians and local authorities that highway schemes should be built and that they were an essential part of the modernization process.

There was also a debate between those who were arguing for comprehensive long-term planning with clearly defined levels of investment and those who were promoting a more piecemeal approach to investment. This is best illustrated in the influential papers by Beesley and Kain (1964 and 1965) which criticized the comprehensive approach to transport planning proposed in *Traffic in Towns* (Ministry of Transport, 1963). They argued that the comprehensive approach

> misrepresents the problems attendant upon rising car ownership now, and in the coming decades, and much overstates the extent of the problem and the scale and kinds of policies needed to deal with it. This is unfortunate because, if anything, it *reduces* the probability that more investment will be prised from a reluctant Treasury, which is skilled at spotting flaws in advocacy. In concentrating attention on action necessary to meet a situation held likely to arise in 2010, the Report (*Traffic in Towns*) encourages easy acceptance allied with practical determination to do very little. More important, however, the Report fails to provide an alternative for action, even if, with the best will in the world, the level of investment expenditures available for urban improvement are much less than the Report's authors feel is desirable. (Beesley and Kain, 1964, p. 200)

Beesley and Kain suggest selective investment in existing primary routes near city centres together with some investment in new infrastructure. The expected growth in car ownership and use predicted in *Traffic in Towns* was exaggerated. Growth could be accommodated through limited investment and extensive traffic management and demand management, including parking controls and road pricing. With the benefit of 30 years hindsight, it is still difficult to say which view was 'correct', particularly as the counterfactual situation is unknown.

A further dimension of the rejection of the 1960s approach was the rise of the environmental movement in the early 1970s and the encouragement given by the Skeffington Report to public and pressure groups to get more fully involved in planning issues. The general public had become much more aware of the question as to whether a road should be built, the weaknesses in established analysis and evaluation methodologies, and the presumption that redevelopment and blight were the inevitable consequence of progress.

The opening of Westway in July 1970 epitomized public concerns. This section of elevated motorway in West London passed through communities, very close to housing and seemed typical of the cavalier approach of highway engineers to existing structures (Van Rest, 1987). It was partly as a response to public concern about schemes such as Westway that the Urban Motorways Project Team was set up (1969)

* to examine current policies for fitting major roads into the urban fabric;

* to consider changes that would improve the integration of roads with their environment;

* to examine the consequences of such changes for public policy, statutory powers and administrative procedures.

However, even in 1969, the government reaffirmed its commitment to 'devote a substantial and growing part of the programme to road works in urban areas' (Ministry of Transport, 1969), and the 1970 White Paper on Roads for the Future required an investment programme of £4000m (at 1970 prices) to be spent over 15–20 years. This programme was designed to 'check and then eliminate the congestion on our trunk road system' (Ministry of Transport, 1970). The main objectives were

* to achieve environmental improvements by diverting long distance traffic, and particularly heavy goods vehicles, from a large number of towns and villages, so as to relieve them of the noise, dirt and danger which they suffer at present;

* to complete, by the early 1980s, a comprehensive network of strategic trunk routes to promote economic growth;

* to link the more remote and less prosperous regions with this new national network;

* to ensure that every major city and town with a population of more than 250,000 will be directly connected to the strategic network and that all with a population of more than 80,000 will be within 10 miles of it;

* to design the network so that it serves all major ports and airports;

* to relieve as many historic towns as possible of through trunk road traffic (Cullingworth, 1988).

The early 1970s marked a sudden change of priority brought about by events at all levels of decision-making. At the national level the Urban Motorways Committee (DOE, 1972*b*) recommended integration in the design of roads with the surrounding area and inclusion of the costs of remedial measures as part of the total construction costs of a scheme. The report was published at the same time as a review of the land compensation aspects of road construction, and together they formed the input to the 1973 Land Compensation Act (Starkie, 1982). The second influential report was that of the House of Commons Expenditure Committee on Urban Transport Planning (1973) which reinforced the

TABLE 2.2. Urban transport planning, 1973.

Two main recommendations

(*a*) National policy should be directed towards promoting public transport and discouraging the use of cars for the journey to work in city areas.

(*b*) Department of the Environment should have a positive policy for urban transport, laying down a broad approach which it should ensure that local authorities follow.

Detailed recommendations under four main headings – many have been implemented (about half by 1980 according to the Specialist Adviser).

(*a*) *Promotion of Public Transport* – including investigation of rapid transit systems, introduction of bus lanes and busways, implementation of bus management schemes, relief to taxis for VAT, operating subsidies for public transport and rates rebates for firms which stagger hours.

(*b*) *Traffic Restraint* – including extension of parking meters, higher prices for parking, reductions in long-stay car parks, local authority control over off-street parking, maximim limits on car parking spaces with new commercial developments, keeper liability for parking offences, increases in the levels of fines for serious parking offenders, more widespread use of pedestrian precincts, routes for heavy lorries, the possibility of shifting taxation from the annual licence to petrol tax and a review of all road schemes in urban areas that had not been actually contracted out.

(*c*) *Transportation Studies* – should still be carried out but with analysis of how people can be deterred from travelling and how many people might make use of improved public transport services. Alternative proposals should be significantly different from each other and the proportion of existing schemes considered as firm commitments should be minimized. The assumptions made and methods used should be more comprehensible to the layman.

(*d*) *Organization and Financing of Transport* – clearer allocation of transport planning responsibilities required for DOE. The urban roads programme should be replaced by an urban transport programme with greater emphasis being given to social research and analysis of public attitudes. New grants should be available for buses and expenditure programmes for urban transport should be based on definable and explicitly stated planning objectives.

*Source*: House of Commons Expenditure Committee (1973).

general trend away from investment in roads to better management and use of existing resources (table 2.2).

With respect to urban transport planning this report marked a watershed as it summarized many of the concerns expressed over the 1960s approach to the production of traffic plans. These traffic plans had been used to test traffic and transport alternatives within a single set of land-use patterns. Often the transport proposals tested were very similar to each other, and distributional issues were not investigated. The Select Committee recommended that transport studies should be carried out within a clearly defined framework of general policy. The bias towards roads investment and motorway construction was evident, and local planning authorities seemed prepared to accept the recommendations of the transport studies without a critical examination of the assumptions.

The main weaknesses identified were the inability to predict how many people will be deterred from travelling under certain restrictions and the inadequacy in predicting how many people might use new or improved public transport facilities. In most transport studies a large proportion of the budget had already been committed to urban motorway projects and many of these schemes had not been fully evaluated. In future a wider range of alternatives should be evaluated and the proportion of schemes considered as firm commitments should be kept to a minimum. Finally, it was proposed that alternative strategies should be considered at an early stage, and that the assumptions made and the methods used in transport studies should be comprehensible.

The recommendations made by the Select Committee summarized the growing dissatisfaction with the approach adopted and the priorities allocated to road construction in the 1960s. By capturing the public, the academic and the political concerns, it managed to achieve a switch in priorities away from road plans to the better management and use of the existing infrastructure through restraint on the use of the private car and through priority to public transport. In terms of analysis methods it stressed the importance of evaluation, the strong links between land use and transport, and the issues which were of concern to policy-makers, in particular the modal choice decision.

At the local level political opposition to urban road building was growing. The most celebrated case outside London was the decision in Nottingham to abandon much of their highway construction plans and to look at the possibility of controlling traffic entering the city centre. The zone and collar scheme was introduced in the western part of the city on an experimental basis in 1975. Private cars would have limited access to the city centre and would be delayed by traffic lights at peak periods. Long-stay city centre car parking was reduced and prices raised. The intention was to get car users to switch to public transport for city centre trips and to use park and ride schemes. The Nottingham scheme was not a success and was abandoned shortly afterwards (1976). However, its significance was that it reflected public concern over urban road building and it demonstrated that an alternative did exist. However, it also illustrated the difficulty of introducing radical transport policies, particularly those which required people to switch from car to public transport. Getting people out of their cars has proved to be one of the hardest and most politically sensitive transport policy objectives. In retrospect, it can be seen that it was unwise to use time delay as a means to control traffic. Time is a resource which cannot be saved, so inefficiency was introduced into the system unnecessarily. It may have been preferable to have used a pricing mechanism to limit access, but this would have created its own problems, namely political acceptability.

In London, the London Amenity and Transport Association (LATA) produced its own report on Motorways in London (Thomson, 1969) which made a strong case against any urban motorways in the capital. This report coincided with the Greater London Development Plan inquiry set up to consider objections to that Plan. These objections exceeded 2,800 and mainly related to the transport proposals for a series of orbital motorways. The Panel recommended scaling down the orbitals to the M25 (substantially outside the GLC area) and the most controversial inner ring road. However, it was the political changes in London that marked the end of these urban road schemes. The London Labour Party, after much debate, decided to fight the 1973 elections on an anti-motorway platform. This decision won them power and London's motorway plans were effectively abandoned for the next ten years.

Although some cities, such as Glasgow, Leeds and (as noted earlier) Liverpool, maintained urban roads programmes, most cities abandoned their road building programmes, at least temporarily. The philosophy of the Buchanan Report which had been the standard for policy in the previous decade was also abandoned. Cities would no longer be subjected to substantial road building programmes to allow unrestricted access by private cars. In the Autumn of 1973, John Peyton, the Minister of Transport, announced in a short speech to the House of Commons 'that he proposed a switch in resources in the transport sector away from urban road construction.' This statement was made exactly ten years after the publication of *Traffic in Towns* (Starkie, 1982).

The early 1970s marked the end of the aggregate large-scale land use transport model. The doubling of world oil prices in 1973 and the impact that this had on world recession, inflation and unemployment, may have been the trigger to end large-scale urban road investment programmes. However, as noted here, this disenchantment was much wider spread, ranging from the political concerns over the costs and electoral implications of unpopular programmes, through technical concerns over the accuracy and validity of the systems approach, to the social concerns over who was actually benefitting from road building. All these events taken together meant that 1973 marked the most important watershed in thinking on urban transport planning.

## The Re-emergence of the Land-Use Transport Study: The 1970s

It may seem surprising after such a wholesale condemnation of the systems analysis approach to transport planning and its political and public rejection that there was a re-emergence of the approach as part of the new strategic planning system set up in the 1970s. Investment would no longer be argued on the basis of where to allocate economic growth, but that investment was now essential to prevent inner city decline. Modern cities in the late twentieth century required high quality road and public transport infrastructure for their survival.

The 1968 Transport Act set up four Passenger Transport Authorities in the main conurbations (Merseyside, the South East Lancashire and North East Cheshire [SELNEC] area, the West Midlands, and Tyneside) to integrate all forms of transport and to coordinate transport with urban planning. West Yorkshire and South Yorkshire became two new PTAs after the 1972 Local Government Act. With the restructuring of local government all county authorities were required to draw up Transport Policies and Programmes (operational from 1974) in which comprehensive policies for the development and operation of transport were presented. These annual plans were to cover roads, traffic management, parking and public transport, and their content should include a statement on local transport objectives, estimates of income and expenditure for the following financial year and a programme to implement the stated longer-term (10–15 years) strategy. The Department of the Environment (and the Department of Transport from 1976) would judge whether the preparation of the TPPs was adequate and the overall costs were within available resources. It would also ensure national and regional considerations were taken into account. The TPP was the basis on which support for transport from central government was distributed to the counties through the Rate Support Grant and the Transport Supplementary Grant (TSG was only about 4 per cent of the size of the RSG). Local government reorganization, together with the new TPP system, was designed to replace the plethora of different authorities and agencies with less than 50 counties which would now be responsible for transport planning. As Starkie (1973) summarizes, these authorities would be adequate 'both in size and in resources to operate with a sensible external independence and internal cohesion, thus providing a stable environment for policies to take effect.' The specific transport requirements formed part of local government reorganization which also set up a system of structure and local plans. The TPPs should be linked in with the structure plan strategy and take account of the detailed implementation of policy outlined in the local plans.

Returning to Merseyside as a case study, the Structure Plan team did not attempt a systematic analysis for the options as had been carried out in the 1960s MALTS study. Instead, they took a wider perspective on the future of the conurbation and one based on much less data collection and analysis, but with little emphasis on the integration between transport and land-use planning. Although, on the land-use side, the Structure Plan's concern over inner-city regeneration might improve the prospects for public transport, nothing was said about how an increase in revenue support for public transport might be expected to make a positive contribution to that regeneration.

The influence of MALTS was still in evidence. Even though over 50 of the original roads schemes were abandoned, 17 were retained in the preparation pool and 10 were approved for construction before 1986. But more striking was the way in which they were justified. While some of the same arguments used 15 years earlier were still put forward, essentially the schemes were now seen as a means to promote economic development and traffic growth, not as a means to accommodate growth that was assumed to be inevitable.

Despite all the changes that had taken place in the planning assumptions, the inner ring road in Liverpool was reinstated. It received a tremendous battering at the examination in public of the Structure Plan, and was finally axed by the Labour administration which took over Liverpool in 1981. It seems that the same approach adopted in the MALTS study had re-emerged in the 1970s in a slightly modified form.

An examination of all conurbations would reveal many features in common with Merseyside. In the 1960s and early 1970s, large-scale aggregate models were used. In the late 1970s and early 1980s these gave way to a broader based approach to planning. At the same time, a policy based on the allocation of growth through dispersal had been replaced by one aimed at urban regeneration. Most remarkable is the apparent continuity of some of the key elements, highlighted in the Merseyside evidence (Banister and Botham, 1985).

- It took nearly 20 years to realize that city centre congestion was not a problem and that the present infrastructure was more than adequate.

- Transport policies were still thought to be effective instruments of land-use policy. In the 1960s that faith was attached to road building, but in the 1970s it was attached to low fares.

- Methods of analysis had not improved. The MALTS methodology made what can be seen to be serious mistakes, because the assumptions it was built on turned out to be wrong. The Structure Plan process which followed it aimed to avoid what was seen as the inflexibility of the old style plan. But in the end it undertook no analysis to justify those transport policies which were being promoted.

During the 1970s the re-emergence of transport planning was a continuation of the 1960s approach, but under the guise of the Structure Plan and to a lesser extent the Transport Policy and Programmes. The analytical skills were still required, but more in the planning context, to predict changes in population, employment and local economic activity (Batty, 1976). However, it was during the 1970s that several other transport policy issues required the attention of researchers. The common factor underlying each issue was the new policy imperative to ensure value for money in times of reduced public investment, and to make the best possible use of existing infrastructure and resources. These concerns manifest themselves in a variety of ways, five of which will be highlighted here.

DEMAND MANAGEMENT

Demand Management required the best use of the existing infrastructure through traffic control, bus priority schemes and the regulation of the supply and price of public car parking space. The government's Transport and Road Research Laboratory carried out a series of experiments on urban traffic control including the Leicester (1974) purpose built automated traffic control system. The adoption of bus lanes was less than enthusiastic in the early 1970s and parking controls seemed to lack effectiveness as the costs of collection were 50 per cent higher than the revenue recovered in 1972/73 (costs of £48 m and revenues of £32 m). The most radical proposal was the feasibility study carried out in 1973 on area licensing in central London where a supplementary charge would be made on cars and other vehicles entering the licensing area. In 1974 the technical officers reported favourably but the new Labour GLC decided not to take the scheme any further.

## THE ROADS PROGRAMME

The Roads Programme in the 1970s was severely curtailed as the government increasingly emphasized coordination in transport planning through the TPP circulars, and this imperative was reinforced by a review of all major road schemes in each county's programme which were not due to start before April 1978 (Circular 125/75, Department of the Environment, 1975). The government's expenditure cuts in the 1970s fell heavily on the roads programme. When the White Paper on Transport Policy was published in 1977 (Department of Transport, 1977a) it was revealed that road construction in progress on the English motorway and trunk roads programme had contracted from about £400 m in 1971/72 to about £270 m in 1976/77 at 1970 prices. Thus new trunk road building had contracted by 32 per cent but local authority road building by 50 per cent (Van Rest, 1987).

Many of the ambitious highway plans recommended by the large-scale aggregate transport studies of the 1960s and early 1970s were axed. Most of them were behind schedule, and economic recession and the restructuring of industry had led to a decline in inner-city employment and pressure on peripheral development, often on sites made accessible by new motorways. The roads that were actually built seemed to fit in more closely with other priorities such as comprehensive traffic management schemes.

## THE ENVIRONMENT

The Environment was also an issue which had raised public concerns, both in terms of the increased land take required for new roads and urban development, and the externalities created by noise and pollution from road traffic. The benefits of a high speed motorway and primary road system for the user were being balanced against the loss of land and a general reduction in the quality of the environment for others. A series of reviews and measures were taken, principally against the lorry, through the Armitage Inquiry (Department of Transport, 1980) and the Foster Inquiry (Department of Transport, 1979*b*). The size of lorries proved to be one issue upon which the environmental lobbies influenced policy. For a decade they were successful in delaying and blocking the powerful industrial interests who wanted an increase in the gross weight and dimension of lorries (Starkie, 1982). It was only at the end of the decade that the argument for heavier and larger lorries was won, and even then it seemed that this victory was because of recession and low economic growth brought about by the second oil crisis in 1979 and not the actual merits of the case. In times of economic recession and hardships, environmental issues become of secondary importance. A lesson which may still be true a decade later!

The government proposed that the resolution to the lorry problem would be the construction of a lorry network, but this proposal in turn became a non-event as the whole roads programme came to a halt when public expenditure was cut back. However, local authorities did have powers to restrict heavy lorry traffic (over 3 tonnes unladen weight) to particular routes 'so as to preserve or improve the amenities of their areas' (Heavy Commercial Vehicle (Controls and Regulations) Act 1973). In both the cases of lorry weights and sizes and of lorry routes, the arguments between reductions in hauliers' costs and environmental benefits lay at the root of the disagreements between the industry and environmental lobbyists. More fundamentally, the 1970s marked a transition in transport planning analysis. No longer was the analytical expertise of the technocrat seen as unquestionable. There were as many experts arguing against a scheme as for a scheme, and for the first time evidence, which had previously been accepted, was now being placed under the microscope.

## THE PUBLIC INQUIRY

The Public Inquiry was the forum within which many of these debates were held. As with the rejection of the 1960s approach, concern with the process of consultation and participation was at first muted, but soon it became a full scale confrontation between the opposing interests. There were several reasons for this discontent. The government maintained that the public inquiry could not question public policy on roads or the basis on which traffic forecasts were made to justify the schemes. The protestors claimed that they should be able to question the need for a road, and they were also concerned about the independence of the inspector (appointed by the Secretary of State for the Environment and Transport) and the treatment of those protestors with legitimate cases at the inquiry, including the availability of information to them and the costs of being represented at an inquiry. Their success was not so much in getting decisions reversed, but in raising public concerns over roads, in gaining publicity for their cause, and in delaying the road building programme. It was now taking over 12 years for a road to complete the process of preparation, inquiry and construction.

Two major initiatives were set up by the government in the mid-1970s. In December 1976, the Advisory Committee on Trunk Road Assessment was set up to review the Department of Transport's Trunk Road Assessment methods, including the procedures used for traffic forecasts and the relative importance attached to economic and environmental factors (see pages 57–60). The second inquiry was made into the procedures surrounding the conduct of public inquiries into highway proposals (Department of Transport, 1978a, 1978b). Certain changes were recommended including the appointment of inspectors by the Lord Chancellor rather than the government department, and the introduction of a pre-inquiry procedural stage which would allow all information to be pooled and permit the basic programme of the inquiry to be agreed. Objectors were still not allowed to question government policy, but they could debate whether national forecasts were appropriate to that particular local situation or were there factors which might reduce or enhance the national forecasts.

It should be noted that statutory objectors who were likely to be materially affected by the proposed road scheme had a right to be represented at the inquiry, but that other objectors could only attend at the inspector's discretion. This distinction between the two basic types of objectors had been introduced in July 1976 just after the review of the public inquiries into highway proposals was set up. These two actions effectively diffused much of the opposition to the roads proposals, but, as noted earlier, the roads programme itself had been severely reduced through the public expenditure cutbacks of the 1970s. In future the government would also publish its roads programme as part of the new openness on information (Department of Transport, 1978a). Gradually, the confrontation between the various interest groups in challenging government policy and forecasts at public inquiries was dissipated. Priorities for road investment were the completion of the primary network of long distance motorways and a switch of funding to small town bypasses. The Policy for Roads document was published regularly and a layman's guide to the role of public inquiries into road proposals was also produced (1981). The temperature of the debate was again raised in February 1980 when the Law Lords overruled Lord Denning's ruling that objectors could cross-examine official witnesses on how they had calculated traffic forecasts. The Law Lords ruled that local inquiries were not the appropriate place to challenge the forecasts.

The government had been extremely successful in dissipating many of the legitimate concerns of the public through the introduction of a more sensitive and open procedure for the public inquiry, but probably the main reason was the end of the 1970s investment programme in new motorways. The rationale for protest had disappeared.

PUBLIC TRANSPORT

Public Transport patronage during the second part of the 1970s continued to decline with the bus now accounting for 12 per cent of travel (Department of Transport, 1976a) and an increasing level of public subsidy. Total support to the bus industry more than doubled in just two years to 1975/76, with the bulk of subsidy going to metropolitan areas. It seemed that the difficulty of getting people to use public transport in preference to their car was really an impossibility. Public transport was increasingly only being used by those people without access to a car and the laudable aim of getting people to switch modes was unrealistic. The questions posed by the House of Commons Expenditure Committee (1973) and table 2.2 have been answered, namely that there are no methods available to predict modal shift back to the bus unless a comprehensive approach to planning is adopted to give buses exclusive rights of way in cities. It is not just a question of generalized cost differences between the alternative modes, but the quality of the service offered by the bus. It has taken another decade of decline fully to realize the implications of this problem.

Even after the 1970s bus travel has continued to decline, with reduced levels of service, increases in costs of operation and increases in fares which rose by 30 per cent in real terms (1972–1982). Revenue support has also grown from £10 million in 1972 to £520 million in 1982, a thirteen fold increase in real terms (Department of Transport, 1984). Many conurbations had increased their support for public transport, mainly through maintaining a low fares policy. In Merseyside, ridership has increased from a low point of 216 million bus trips in 1981/82 to 254 million in 1984/85, with revenue support for buses rising from £11.8 million in 1981/82 to £18.2 million in 1984/85 (at 1975 prices). In an attempt to defend their position, the County's 1985–88 TPP submission quoted the Structure Plan panel to emphasize its case

> we believe that an effective public transport system will contribute to regeneration because it will broaden the range of accessible job opportunities, especially for the less well off, and will extend the catchment area within which an enterprise on Merseyside can recruit . . . Our opinion is that failure to maintain an effective public transport system will hinder the fulfilment of the strategy and will do so more grievously than a lack of resources to carry through new road construction.

The significance of being able to quote the Structure Plan and report of the panel presiding over the examination in public is that the government had accepted the plan in 1980. The County's submission went on to note that

> the value of this turning of the tide (in patronage) for the urban regeneration strategy needs to be measured in terms of the social, economic and environmental needs addressed in the Structure Plan's policies.

But the County Council had made little attempt to quantify the projected impact of cheap fares, nor did it monitor their effect on urban regeneration, once they were introduced. Thus, despite all the claims put forward for the policy, including its redistributive effects, there has been little attempt or empirical evidence to justify any of them.

The Merseyside policy outlined here was stated in the mid-1980s. This in itself is strange as the Conservative government had already made it clear in a series of Transport Acts (1980, 1981, 1982, 1983) that public expenditure would be reduced and that subsidy levels could not continue to increase. The thinking in Merseyside was more appropriate to 1975 when the Labour government was preparing its major piece of legislation on a social transport policy (the 1978 Transport Act).

This major statement of transport policy (Department of Transport 1976*a*, 1977*a*) identified three policy objectives:

- To contribute to economic growth and higher national prosperity, particularly through providing an efficient service to industry, commerce and agriculture.

- To meet social needs by securing a reasonable level of personal mobility, in particular by maintaining public transport for the many people who do not have the effective choice of travelling by car.

- To minimize the harmful effects, in loss of life and damage to the environment, that are the direct physical result of the transport we use.

Other secondary objectives emphasized the efficient use of resources, notably energy, the importance of choice and local democracy, the interests of transport workers, and the need to restrain public expenditure. The well-established objective of providing an efficient transport system, that would contribute to economic growth and higher national prosperity, was supplemented by the social objective of meeting transport needs.

The county councils were also given stronger powers to coordinate transport and a requirement to prepare a County Public Transport Plan (PTP) to cover public transport within the context of the TPP. The Plan should set out the policy for public transport in the county for five years ahead, and to assess the transport needs of those people in the county and how those needs could be met.

Towards the end of the 1970s a considerable amount of research took place to determine how needs could be measured. This included an assessment of the different approaches adopted by the Shire Counties (Banister, 1981). Three basic methods were used – minimum levels of service or standards, demand measures, and local-based measures – but none was satisfactory either theoretically or empirically, and an operational tool for transport planning was still sought. The concentration on the identification and measurement of need may be misplaced, and needs can only be considered at the conceptual level with the aim of providing a perspective within which policy-makers can consider the distributional effects of particular decisions (Banister, 1983). The question of transport needs was never satisfactorily resolved, and with the switch from distributional concerns over needs to those of identifying markets for transport, economic concepts of demand and willingness to pay again came to precedence in the 1980s.

## Conclusions

During the 1960s and 1970s nearly 2300 kilometres of motorways and 2500 kilometres of dual carriageway all-purpose trunk roads had either been constructed or upgraded, and the strategic network was near completion.

In future, priority would be given to specific projects, such as the M25 route round London, and it was recognized at last that transport planning should be more closely related to land-use planning. The assumption of a continued supply of relatively cheap fuel was no longer realistic, and decisions should now be made that integrated land-use functions, so that the absolute dependence on transport and the length of numbers of journeys could be reduced.

Over these two decades there seems to have been a continuity of roads policy (table 2.3) which was only terminated with the 1978 Transport Act when social objectives and commitment to public transport were added to policy concerns over the efficiency of the transport system. Even then this 'end' of road building was only temporary as the radical Conservative governments of the 1980s changed both the focus of transport policy and the frequency with which major policy changes took place. Trends in the 1960s and 1970s in the economic and social cycles were fairly smooth with more abrupt changes brought about by the two 'oil crises' of the 1970s. But transport policy demonstrated 'distinct, quite sudden but infrequent change' (Starkie, 1982).

These changes usually took place towards the end of a particular administration and the continuity was supplied by the next government. Starkie (1982)

TABLE 2.3. Continuity of roads policy in the 1960s and 1970s.

| 1946 | Labour | Adopted the 'tea-room' plan for trunk roads |
|------|--------|---------------------------------------------|
| 1951 | Conservative | Continued with that plan |
| | | Developed urban road traffic policies |
| 1964 | Labour | Continued with the 'Marples' policies |
| | | Revitalized the trunk roads programme |
| 1970 | Conservative | Implemented this strategy |
| | | Developed urban transport policies |
| 1974 | Labour | Adopted the urban policies |
| | | Reappraisal of trunk roads policy |
| 1979 | Conservative | New generation of motorway construction |

*Note*: The tea-room plan was produced in 1946 by Alfred Barnes, the Minister of Transport. It is said that these plans were exhibited in the Member's tea-room at the House of Commons and that the plan looked like an 'hour glass' shaped network of motorways covering some 800 miles, and it was the intention of the Labour government to complete this plan in 10 years (it was only in 1971 that the first 800 miles of motorway were opened in Great Britain).

*Source*: Based on Starkie (1982) and British Road Federation (1972).

argues that the civil servants wished to avoid conflict and seek consensus, and this strategy deflected challenges to existing policy. If a sufficient consensus against existing policy fails to focus upon a generally agreed new course of action, policy regresses towards a non-policy (Plowden, 1973), such as the switch away from urban motorways to managing demand.

*Chapter 3*

# Developments in Planning Analysis and Evaluation

## Parallels in Land-Use Analysis

Over a similar period of time, an almost parallel development of analysis processes was taking place in land-use and urban systems analysis. Again there seemed to be a natural progression from the realization of the complexity of urban problems and the use of the scientific method for the analysis of social problems towards a systems view of the way in which cities and regions were being perceived. Throughout their development, land-use models have been linked directly to public policy, hence the precarious nature of their existence. There is a compromise between the theoretical acceptability and their practical feasibility (Batty, 1981).

The years 1960 to 1965 saw an explosion in land-use modelling. These early models were based on fairly strong assumptions and were concerned with the representation of the city. As such they were aggregate in nature and vast in scale. The expectations were that modellers and planners would produce tighter and stronger answers to urban problems, but in reality there was a mismatch between these expectations and the realization of the vast size of the task. At the end of the 1960s there was a fairly general feeling that the models had been oversold and the initial optimism was fading (Butler *et al.*, 1969).

Three particular problems seemed to summarize the difficulties. Many of the models were never actually built because of their size and complexity. Secondly, the models did not explain phenomena of interest to decision-makers and this was also reflected in the ludicrous predictions of growth in urban areas. Finally, the organizational environment was not conducive to the development work necessary to make the models operational. The analyst often had to learn on the job about the problems and there were long periods with no apparent progress. The decision-makers wanted instant answers from instant models. More fundamental though was the absence of any coherent theory of how urban systems worked, and the methods themselves had often been taken from other disciplines, in particular the physical sciences, operations research and statistics.

The gentle critique soon developed into a full scale review. It was started by Boyce, Day and McDonald's review (1970) of metropolitan plan-making in thirteen major urban areas in the USA. They examined the approaches adopted by planning teams in preparing and evaluating alternative strategies, and found that there had been too many abortive data collections on a massive scale, that too much time had been spent on the development of techniques and not enough time on the evaluation of alternatives, and that in most cases the planning process was conceptualized as a linear sequence leading to the selection of one plan from a range of alternatives. There were also more general attacks on the whole field of systems analysis and its use in public policy (Hoos, 1972; Berlinski, 1976) and on particular applications of land use models. For example, Brewer (1973) reviewed the mismanagement of the San Francisco and Pittsburgh Community Renewal Programmes, both in terms of the methods used and the organization of the studies themselves. The culmination of the review was Lee's (1973) requiem for large-scale models (see pages 29–30).

In almost every case it seemed that the time, the data and the expertise were not available to develop a general land-use and transport model which was consistent in its level of detail. The only exception was Lowry's model developed for the Pittsburgh Regional Economic Study (Lowry, 1964). This model simulated urban development, given information about the spatial distribution of basic employment and the costs of travel. The model never achieved popularity in the USA, but was used in the early sub-regional studies in Britain (Batey and Breheny, 1978).

The 1970s marked the revival of systems analysis, with more theoretical work and more modest claims for what they could and could not achieve. Pack's review (1978) of work in the USA demonstrates that there was a proliferation of modelling applications in the 1970s. The use of more transparent methods, better communication and the breaking down of barriers between planners and decision-makers all facilitated this process. One of the most effective arguments put forward by Kain (1978) was that most of the critique had been made against some normative ideal. Effective criticism must be based on alternative ways of tackling urban problems. He cited the Community Renewal Projects in the USA, commenting that there were several hundred of these, but it was only those in Pittsburgh and San Francisco that had any influence on housing research and policy. These were the two which had used land-use modelling techniques.

Land-use activity allocation models capable of estimating future patterns of land development and transport flows were the main contribution of planners. Much of the thinking was based on the Lowry Model (Lowry, 1964) which split employment into two categories, with basic employment being allocated exogenously and service employment being allocated by the model to reflect the distribution of households and employment, both of which were calculated within the model. Other similar approaches all used the gravity type allocation model to link the mechanisms of urban growth and change (Hutchinson, 1981).

Models which calculate urban activity allocations are very dependent upon the land development constraints established exogenously. The main determinant of the levels of interaction is the cost deterrence function, and once the existing activity has been reproduced, the assumption that this function remains stable is crucial to future activity allocation. The impact of changes in the land market through land availability and rent changes, and the advantages of different activities locating in close proximity to each other (agglomeration economies) have not been fully developed.

In Britain, the motivation was similar to that in the USA but the underlying structure of planning was different. Up until the 1960s planning in Britain was dominated by architects and physical planners. It was only at the end of the 1960s that a fundamental change took place with the introduction of social sciences to planning and the adoption of the systems approach to analysis (McLoughlin, 1969). The first urban models were built in Leicester, Manchester, Bedford and Cambridge and they were based on shopping models applied in Haydock, Oxford and Lewisham (Foot, 1981). The links between land use and transport were not as strong as in the USA, and most were built on the sub-regional scale. This first generation of models was designed to allocate growth and the analysis was often carried out by small teams working outside the constraints of the local authority. Outputs were specific with the emphasis on order rather than the US tendency for comprehensiveness and size. Systems analysis was also concerned with comprehensiveness and the rational planning process. Planning methods such as Lichfield's Planning Balance Sheet Appraisal and Hill's Goals Achievement Matrix were developed for evaluation and to supplement the analytical methods such as gravity models, economic base techniques, Lowry models, potential surface models and cohort survival methods (Lichfield, 1970; Hills, 1973).

The growth in the popularity of the systems approach coincided with the reorganization of local government (1974) and the introduction of structure and local plans. In theory this new environment should have marked an increase in the use of models, but the reverse occurred. Structure plans were established in the 1968 Town and Country Planning Act to replace the existing system of Development Plans which were mainly zoning exercises. The new system introduced two tiers of plans (table 3.1). At the strategic level, the structure plans were designed to state the broad policy objectives for the counties, with local plans concerned with the detailed implementation of policy, in particular development control. On the whole the structure plans, with a few notable exceptions (South Hampshire and Teeside) failed to

TABLE 3.1. Development plans to 1986.

---

*Major changes to the Planning System introduced by the Town and Country Planning Act (1968)*

---

*Structure Plans*: Produced by Metropolitan Counties and Shire Counties for Ministerial approval. Statement of a broad strategy for development including associated transport systems and broader social and economic objectives.

The last of the 82 first generation structure plans in England and Wales was approved in July 1985. By April 1986 there were three replacement structure plans and 30 alterations to original plans which had been submitted but not yet approved.

It took 14 years to complete the first cycle. Between 1981–1985 there were 32 approvals with an average time of 28 months from submission to approval. The delay was even longer if public participation and preparation time were included. These very lengthy documents were often out of date when approved.

---

*Local Plans*: Produced by local authorities for adoption. Formulated detailed proposals for the development of land including maps. Set out development control policies either for a limited area of comprehensive change (action areas) or for larger localities (district plans). Subject plans dealt with specific policy issues such as housing, minerals or conservation.

By March 1986, 474 local plans had been adopted and a further 269 were on deposit prior to adoption. It was claimed that many were too detailed and that delays in approval of structure plans had affected local plans.

---

achieve the level of technical sophistication evident in the sub regional plans developed in the late 1960s and early 1970s (Batey and Breheny, 1978).

Up to the 1970s the accepted means of plan-making had been to survey the area, to analyse its known or anticipated problems, and to produce a plan to describe a state of affairs expected or desired at some future date – the classic survey, analysis, plan sequence. The new system of plan-making accepted a process that was continuous, cyclical and took a systems analysis view based on the identification of needs, the formulation of goals and the evaluation of alternatives, with subsequent monitoring and modification of adopted plans.

Again, as in the USA, there was a reaction against systems analysis in Britain. It has been suggested by Batty (1976) that this was part of the action-reaction cycle for and against models and the use of models in planning. In 1967 land-use models were first used in Britain and optimism was high in 1970 when the Department of the Environment issued guidance on the use of predictive models in structure plans. By 1974 the reaction was already setting in with further Department of the Environment instructions (DOE Circular 98/74) suggesting that counties should concentrate on key issues in their structure plans. As with the transport models, there was an increased interest in the economic and social aspects of planning, and a growing scepticism about the ability of planners to produce long-term physical plans. Breheny and

Roberts (1981) concluded that much forecasting work was weak, both in terms of the way forecasts were used and the way they were produced. In particular, their review of 13 of the 60 available structure plans criticized the weakness between forecasting and policy formulation.

The mid-1970s coincided with a downturn in anticipated growth levels as an outcome of the oil crisis (1973–1974) and a move away from the rational comprehensive approach to plan-making towards a pragmatic incremental approach (for example East Sussex). With the introduction of the structure planning system there had been a separation of powers between the new local government counties and the districts. Implementation was at the local level and so structure plans were seen to be largely ineffective, particularly as the process of submission, inquiry and adoption was often lengthy. With the switch from growth to decline, planners had a much harder task. Allocation of growth may be difficult but accommodating decline is much harder. Urban modelling had not suggested that inner-city decline would take place together with the massive decentralization of activities and changes in the employment structure. The priorities for the planning process had changed from forecasting growth towards an emphasis on the short term and the need for flexibility.

Towards the end of the 1970s, planning was changing with the move away from large-scale publicly funded urban renewal schemes towards giving additional powers to local authorities with severe inner-city problems so that they might participate more effectively in local economic development. The Inner Urban Areas Act (1978) set up partnerships (7) in the most severely deprived areas, together with programme authorities (23) and other designated areas (16) in locations with less severe problems. An inner-area programme was to be produced as a bid for funding, and it was to involve all parties (private and public, statutory and voluntary).

This move towards a greater involvement of the private sector was continued after the General Election of 1979 with an extensive urban regeneration programme. Planners were becoming negotiators between the private and public sectors, and their role was no longer primarily regulatory but more development based, seeking opportunities for planning gain (Savitch, 1988). The Urban Development Corporation was the main means by which property investment could be stimulated. The responsibilities of the local authority planners were pre-empted as the role of the UDCs was to make sites available to developers and to provide the appropriate infrastructure so that investment would follow. It was ironic that one of the most interventionist actions concerning urban planning was taken by a government which was supposed to allow the market to operate (Lawless, 1986).

The 1980s have provided a major paradox in UK planning, with strategic planning becoming increasingly unfashionable in both central and local government, yet many strategic planning issues have captured the headlines (Breheny, 1989). The concern over long-term development and environmental quality had been replaced by short-term gains and negotiations. The new definition of planning was the process by which government enables the private sector to invest profitably in urban areas. The earlier construction of urban problems as defined by poverty and inner-city decline was reformulated in terms of competitiveness and fiscal solvency (Smith, 1988). The government's intentions were made clear in the White Paper 'Streamlining the Cities' (1983) where it was stated that the earlier confidence in strategic planning has been 'exaggerated' (para 1.3) and the search for such a strategic role may have little basis 'in real needs' (para 1.11). The abolition of the strategic level of government in London and the Metropolitan Counties took place in 1986 (table 3.2). This was followed by the proposed abolition of structure planning in the Consultation Paper, 'The Future of Development Plans' (September, 1986). This proposal was formalized in the White Paper of the same title (January, 1989), but no action was taken. As Breheny (1989) observes, planning at the strategic level and many of the administrative structures which were designed to support it have been largely dismantled as

TABLE 3.2. Development plans since 1986.

---

*Major changes to the Planning System introduced by the Local Government Act 1985*

---

1. *Strategic Planning Guidance* from the Secretary of State which was not formalized or centrally controlled but worked out in conjunction with local planning authorities with a draft for public comment.

2. *Strategic Policies* to be established by the Counties to guide the detailed preparation of Development Plans by the district authorities. The Strategic Policy to cover

- strategic highways and transport
- policies on control of mineral operations and waste
- provision of land for housing
- policies on major retailing and industrial development
- protection of the countryside

The County responsibilities also covered survey, analysis and the preparation of information, the process of consultation before publishing draft statements, the provision of an examination in public after consultation, and the publication of the adopted statement. There was no longer a requirement for the Secretary of State's approval.

3. *Unitary Development Plans* (for London Boroughs and Metropolitan District councils) *and District Development Plans*. These detailed documents are similar to local plans and cover all land-use policies (except minerals). They identify proposed locations for development.

Draft plans are published with a period of six weeks for representations and objections together with consultations with Counties and the government. A public local inquiry would be held with a planning inspector to give an independent commentary on the plan. A plan is then published with modifications and adoption takes place after a further six-week period for comment. The adopted plan does not require the Secretary of State's approval, but he would receive a copy.

---

part of the continuing process of streamlining the administrative procedures surrounding land-use planning.

The new pragmatism and highly centralized planning process replaced the lengthy and data intensive strategic planning process by brief regional guidance notes from government, supplemented by strategic policy statements produced by the counties (table 3.3). As early as 1986, the Secretary of State for the Environment announced that structure plans would be abolished as they were 'unwieldy and took too long to prepare'. It was only in 1989 that the White Paper proposed the replacement of structure plans by county statements which would give a broad outline of policy with details to be filled in by district development plans. These proposals have not been implemented but structure plans now no longer need Department of the Environment approval. They still have to go through public consultation and examination in public, and there is a requirement to cut out detail and to concentrate on key strategic decisions.

It is also ironic that as strategic planning disappears from public sector planning it is making an appearance in the private sector. Many large corporations are now committed to forward planning to assist with more rigorous decision-making (Breheny, 1991). This interest is particularly apparent in the retail sector where several of the major companies have set up specialist groups or employed consultants to carry out systematic analysis to identify suitable locations for new outlets and to assess the options for new forms of distribution. Many of the techniques developed in strategic planning (for example gravity models and operations research techniques) are being used. In addition, many private corporations would like to operate within a strategically planned environment as this reduces uncertainty and hence their own risks. Although many businesses may support the principles of free competition,

TABLE 3.3 The Patten Model.[1]

| | |
|---|---|
| *Regional Scale:* | County authorities form themselves into appropriate regional groupings to produce statements of regional advice (10–15 year period) which is then submitted to the Secretary of State for the Environment. Regional guidance for each region is then produced. |
| *Structure Plans:* | Structure plans will continue to be legislatively based but will be self-approving through consultation processes. The new structure plans will not have to be submitted to the Secretary of State, but there will still be reserve powers to intervene. The new procedures will reduce time for plan preparation and allow more local initiative. |
| *Local Plans:* | No change here except that the plans are now district wide. They are still legally required and must be self approved. |

1. Chris Patten was Secretary of State for the Environment at this time; he is now Governor of Hong Kong.
*Source*: Based on Breheny (1991).

when it comes to their own business most of them do not welcome competition, particularly if it is perceived to be unfair. For all its faults the planning system has provided both continuity and stability, and each entrepreneur was fully aware of the procedures to follow. It is only at the very localized level that planning controls have been retained. At the strategic level for major invest-ment, such as a new settlement or a regional shopping centre, there is less confidence in the new system of limited strategic guidance and the Patten model (Breheny, 1991).

## Evaluation in Planning and Transport

### Cost Benefit Analysis

Across all of transport and planning analysis the issue of evaluation has always proved to be particularly controversial. Ever since Dupuit (1844) concluded that more general benefits accrue to society than is apparent in revenues, decision-makers have been searching for techniques that can include the full range of factors in one analysis. It should be remembered that the basic purpose of evaluation is to assist the decision-makers in choosing between alternative courses of action. Early assessments were made on operational criteria such as whether it was physically possible for roads to accommodate traffic without unreasonable delay. Questions such as design standards, gra-dients and alignments were all important as were the costs, scale and timing of the improvement.

Cost benefit analysis can be defined as 'a practical way of assessing the desirability of projects, where it is important to take a long view (in the sense of looking at repercussions in the further as well as the nearer future) and a wide view (in the sense of allowing for side effects of many kinds on many persons, industries, regions, etc)' (Prest and Turvey, 1965). Social cost benefit analysis goes beyond the consideration of private costs and benefits to the individual enterprise and covers a range of social costs and benefits. As such it draws on concepts from neo-classical economics on the workings of advance capitalist economies. Welfare economics argues that within a general equilib-rium situation, consumers exercise their preferences to consume precisely that combination of goods and services which best suits them. Producers respond to these actions by providing those goods and services which best meet these demands. Through the operation of the capital and labour markets this process is simultaneous with the selection by firms of the optimum technical processes, and the optimum combination of labour and capital inputs. Work-ers can balance their time between work and leisure activities. The distribution of earned income is determined by the interplay of demand and supply, and the distribution of wealth and unearned income is the outcome of savings behaviour and ownership patterns. Welfare economics attempts to accommodate some of the weaknesses of general equilibrium theory as well as extending it to cover some of the income distribution questions. However, many problems still remain such as the treatment of value, market imperfections and the treatment of consumer preferences. Some commentators (for example Dunleavy and Duncan, 1989) dismiss cost benefit analysis as a neo-Keynesian device for boosting public expenditure.

With respect to transport evaluation, cost benefit analysis has been very influential since its first application to the London – Birmingham motorway (Coburn, Beesley and Reynolds, 1960). Losses in travel times and increases in accidents were calculated from the decision not to improve the existing route to motorway standards. This basic approach was superseded in 1972 by the COBA (Cost Benefit Analysis) method of appraisal (DOE, 1972*a*), where the costs and benefits of a new road were discounted over a period of 30 years from the proposed opening date to give a net present value and a net present value cost ratio. It was solely an economic appraisal method involving savings in travel time, savings in vehicle operating costs and lower accident rates, which were matched against the capital and maintenance costs of the new road.

Social cost benefit analysis (SCBA) was first used in transport evaluation for the Victoria line study. In this application, the strict interpretation of user costs and benefits was extended to include the reduction in journey time and costs; the improved comfort and convenience experienced by those travellers diverting to the new line, the time and cost savings to traffic which remained on parallel routes; and the benefits to new generated traffic on the Victoria line (Foster and Beesley, 1963). This broader based evaluation transformed an expected loss of £2.14 million per annum to a healthy rate of return of 11.3 per cent (with a discount rate of 6 per cent), with over half the total benefits accruing to traffic which did not even use the new Victoria line.

By the middle of the 1960s cost benefit analysis was a common prerequisite of project approval with a minimum rate of return of 8 per cent required for public investment. However, it was at the end of the decade that SCBA had its greatest test with the two and a half years of the Roskill Commission Inquiry into the location of London's Third Airport (1968–1970). The scale of the inquiry and the thoroughness of the methods used made this the pinnacle of rational comprehensive planning. It was at this point that reaction against and rejection of this approach to planning and transport analysis took place. This change was most evident in the Greater London Development Plan which followed immediately after the completion of the Roskill Commission Inquiry (1969 with the Inquiry lasting for two years from 1970–1972).

THE ROSKILL COMMISSION INQUIRY INTO LONDON'S THIRD AIRPORT

The Commission started by examining a list of 78 possible sites, and within seven months had reduced this list to four: Thurleigh (near Bedford),

Cublington (in Buckinghamshire), Nuthampstead (in northern Hertfordshire) and Foulness (later known as Maplin, on the Essex coast). Stansted was missing as it had already been excluded as being inferior to Nuthampstead on the grounds of air traffic control, of noise impact, and for poorer surface access. Similarly, Foulness had been included as it ranked best on noise, defence, and air traffic control, but it was a very expensive site to develop and was furthest from London with a high surface access cost. Logically, this option should not have been considered further, but it was. The main part of the Commission's work was in the evaluation of the four shortlisted sites.

This comprehensive evaluation compared the costs of the sites as compared with the cheapest overall alternative (Cublington). It was argued that the decision was an inter site comparison and the absolute values were meaningless in themselves. Thurleigh was about £70 million more expensive than Cublington whilst Nuthampstead and Foulness were respectively £129 million and £158 million more expensive in terms of their full social costs and benefits. More interestingly, it seemed that these differences were dominated by a relatively few factors. Passenger access costs alone ruled out Foulness, as these costs accounted for £131 million of the difference in total costs. Even when sensitivity analysis was carried out, little change was found in the overall results. The team concluded that even if substantially different values of time were used, the rankings would remain unchanged. Access costs to Foulness far outweighed all other costs, including construction costs. The most thorough application of cost benefit analysis to a major transport infrastructure decision generated tremendous debate over the exclusive use of monetary valuations, for its omission of many planning and economic development factors, for its valuation of certain factors (particularly time and noise), and for its assumption that all people valued money identically (a benefit of £1 was valued identically by people in different income groups). The Commission accepted that the decision should not be solely based on the cost benefit analysis, but that it should be modified by judgement.

Despite the undoubted environmental disadvantages the Roskill team opted for Cublington. Sir Colin Buchanan could not agree with his colleagues and argued that Foulness was the only acceptable site despite the higher costs of development. Public sympathy was with him and, in April 1971, the government announced that the third London airport would be built at Maplin Sands (or Foulness) in Essex.

The Roskill Commission Inquiry decision encapsulates all the main problems inherent in SCBA. The first problem was the forecasting of the level of traffic, which in the case of air relates to the growth in income levels, the value of leisure time, whether the Channel Tunnel would be built. The Maplin Review (Department of Trade, 1974) scaled down the Roskill predictions on traffic forecasts of air passengers from 82.7 million in 1985 to a range of 58–76 million for the London area. The picture is further complicated by the assumptions

TABLE 3.4 The Maplin Review summary.

| | *1973 Actual Million Passengers* | *1990 Scenarios to Accommodate Traffic* | | | |
|---|---|---|---|---|---|
| | | I | II | III | IV |
| Heathrow | 20.3 | 38 | 38 | 38 | 53 |
| Gatwick | 5.7 | 16 | 16 | 16 | 25 |
| Stansted | 0.2 | - | 16 | 4 | 4 |
| Luton | 3.2 | - | 10 | 3 | 3 |
| Maplin | - | 28 | - | - | - |
| Provincial airports | - | - | 5 | 24 | - |
| Total | 29.4 | 82[1] | 85 | 85 | 85 |

| *Costs* | *Capacity Increase (passengers/year)* | | *Airport Development Costs* | *Transport Capital* |
|---|---|---|---|---|
| | *from* | *to* | | |
| Heathrow | 38m | 53m | £115 m | £90 m |
| Gatwick | 16m | 25m | £70 m | £60 m |
| Stansted | 1m | 4m | £15 m | nil |
| Luton | 4m | 16m | £110 m | £47 m |
| Maplin | 3m | 10m | £70 m | £10 m |
| Provincial airports | 0m | 28m | £400 m | £235 m |

*Notes:*
1. Lower as Maplin is assumed to attract fewer passengers.
2. Prices at 1974 levels.

*Source*: Department of Trade (1974).

made on loading factors, the size of aircraft and the number of movements possible within the available airspace. It also depended on the number of flights that could be accommodated within the existing London airports (Heathrow, Gatwick, Stansted and Luton). The Maplin Review argued that the determinant factor was not air traffic capacity but terminal capacity, and that the existing four airports could all cope with additional terminals.

In the subsequent debate on the accuracy of forecasts (summarized in table 3.4), the fundamental argument switched away from a comprehensive evaluation of the alternative sites, as carried out by the Roskill Commission, to a much more restricted assessment of whether the expected growth of traffic could be accommodated within existing airports. In addition it meant that the £650 m (1974 prices) development costs for Maplin could be saved. It was therefore not surprising that the decision to build Maplin was abandoned (July 1974). Both the recommendations of the Roskill Commission and the most comprehensive evaluation ever undertaken on a transport investment decision had been rejected in favour of more modest developments at existing airports. In retrospect, this may have been the obvious answer, particularly in light of the uncertainty and global recession created by the oil crisis of 1973–1974.

However, it is ironic that many of the same arguments are again being rehearsed and the Civil Aviation Authority is forecasting a doubling of demand for air travel (1987–2005) with a 13 million shortfall expected in the capacity of the London area airports (table 3.5). It should also be noted that actual demand in 1987 was 60 million passenger movements, whilst the Roskill research team's estimate (1971) was for 82.7 million passengers in 1985 and the Maplin Review team's estimate (1974) was for 85 million passengers in 1990. The actual growth in demand has been a doubling (1973–1987) whilst Roskill's forecast was a 2.8 times increase (1973–1985) and the Maplin review was for 2.9 times increase (1973–1990). Forecasting traffic growth has always been problematic both for road and air traffic, and no where is it better illustrated that in this case of air travel.

A second problem is in the valuation of costs and benefits arising from any investment decision. Welfare economics is a theory of value, and is based on the proposition that under conditions of equilibrium, the consumer gains equal marginal value out of each marginal increase in consumption from the goods and services bought. Since all consumers are buying goods and services at the same prices, the equilibrium prices prevailing are a true measure of value. These very strict conditions do not occur and as there are many market imperfections (for example the assumption of perfect knowledge, perfect competition, being able to set prices at marginal cost levels, and the absence of externalities), necessary avoiding action is taken with the adoption of second best solutions. Yet, in the cost benefit analysis, the exclusive use of monetary values for costs and benefits led to considerable controversy over the use of such methods for evaluation and, despite extensive analysis to support the robustness of the Commission's recommendations, this nagging doubt still remained. In all calculations, surface access time was dominant and it was here that the value of time for both work and leisure was paramount. If this was the case, then other sites such as Stansted which were close to London should have been included. The irony here was that Nuthampstead was included in the shortlisted sites at the expense of Stansted, but was rejected because of its high noise costs and the loss of agricultural land. Stansted had the same advantages of good access and low construction costs, but less of the disadvantages of noise and agricultural land loss.

TABLE 3.5. Forecast demand for air travel, 1987–2005.

|  | 1987 | 2005 |
| --- | --- | --- |
|  | (million passengers) | |
| Heathrow | 36 | 55 |
| Gatwick | 20 | 30 |
| Stansted | 1 | 33 |
| Luton | 3 | 5 |
| Total | 60 | 123 |

*Source*: Civil Aviation Authority Forecasts, February 1989.

Just before the Maplin Review (Department of Trade, 1974), John Heath, who had been Director of the Economic Services Division at the Board of Trade (1964–1970), gave his verdict on the whole debate at a symposium. It is, he says, one of a series of projects with similar characteristics: long delay between the decision and the final implementation, high uncertainty in technological development or in marketing or in both, and large scale. Heath suggests that Britain has made the mistake of going straight for what was thought to be the ultimate solution, without going through the intermediate stages (Hall, 1980). He concluded 'I believe that it is our traditional system of decision making that has let us down in these cases, and that has also landed us with the present controversy over Maplin' (Foster, 1974).

The large-scale public inquiry commission seems to have been a failure and has not lived up to expectations. The facts can seldom settle an issue in themselves and they may lead to restrictive decisions being made at too early a stage. The short list of just four sites illustrates this problem. Secondly, there must be full agreement within the commission itself. Any dissenting view is seen as weakness and allows the decision to be delayed, so that any perceived challenge to government policy can be dissipated. Roskill encapsulated both of these limitations.

## THE LEITCH COMMITTEE'S INQUIRY INTO TRUNK ROAD APPRAISAL

The experience with the debate about where to locate London's Third Airport crystallized many of the difficulties in strategic decision-making where a long-term view was needed, where large amounts of public expenditure were at stake, where the cancellation of the decision would be expensive and where there were important interests vested against any particular decision being made. A similar history could be traced with respect to the roads programme. The public was raising fundamental concerns over road planning, about the forecasts being used, about the valuation of time and accident savings, about the inquiry procedures, and about the need for the actual road. Previously, there had been little debate over the technical expertise of the planner and the boundaries of professionalism had been clearly drawn. However, since the debate over the motorway proposals for London, that expertise was no longer the exclusive right of the professional. The public were both better informed and wanted more information, they were able to employ their own experts and there seemed to be a general feeling in the mid-1970s that there had to be an alternative to building more roads to accommodate the growth in car traffic (Thomson, 1969; Plowden, 1972).

In the 1973 M42 Motorway Inquiry (Coventry to Derby and Nottingham), the Council for the Conservation Society demonstrated that the Inspector could not exclude any objections to the scheme, in particular whether the road should be built at all. Two years later, the Midlands Motorway Action Group tried to present new evidence on traffic forecasts to the M54 Inquiry (Shrewsbury to M6 North of Birmingham). The Road Construction Unit had circulated their forecasts and had stated that this new evidence 'would not be appropriate'. The Inspector ignored the RCU opinion and heard the new evidence. This decision was swiftly followed by a revised Notes for the Guidance of Panel Inspectors (April 1975) which firmly stated the view that traffic forecasts were a matter for government policy and that consultants or DOE officials should not be questioned on them (Levin, 1979). In 1975 a series of major road inquiries were disrupted and in February 1976 the inquiry into the proposed Airedale trunk road was indefinitely adjourned. The average time for a major road proposal to proceed from inception to completion had now increased from 6–7 years in the 1960s to 10–12 years in the 1970s.

It was partly in response to these criticisms of decision-making on road investments, which were now manifesting themselves in a series of high profile demonstrations at major road public inquiries, that the government set up a comprehensive review of trunk road appraisal chaired by Sir George Leitch (Department of Transport, 1977*b*). This high powered review committee was very critical of existing procedures for forecasting, their insensitivity to policy changes, their unintelligibility, and the overriding priority given to road-user benefits at the expense of other factors, particularly environmental concerns and distributional issues. The review covered existing methods and their evolution, the whole procedure in the development of a trunk road scheme, experience from other countries, a series of worked examples and tests, and a thorough review and critique of existing methodology. It presented a framework for the assessment of trunk road schemes. Its most important contribution was to state systematically many of the public's concerns over analysis and forecasting methods, and it vindicated many of the views held by the pressure groups.

As with the air traffic forecasts for London's Third Airport, *road traffic forecasts* were criticized for not taking account of uncertainties, in particular energy and fuel availability and income growth, and for the way in which single figure forecasts were presented without comment on data or modelling problems. It was also suggested that forecasts tend to be self-fulfilling and tend to reinforce existing trends (Adams, 1981).

The *economic appraisal* itself was criticized for being unintelligible to the layperson, inflexible and an impediment to public participation in the decision-making process. The cost benefit analysis used in the appraisal was considered to be partial, with over 85 per cent of the user benefits being accounted for by savings in travel time. Improved safety accounts for a further 19 per cent and vehicle operating costs about 4 per cent. Vehicle operating costs are a negative benefit because road improvements lead to an increase in these costs as a result of higher speeds and longer journeys (National Audit Office, 1988). The value of time is the most important single factor in the evaluation, and the values used for the calculation of work time and leisure time together with the weight given to small savings in travel times are crucial (Hensher, 1989; MVA *et al.*, 1987). The basic question posed by the Leitch Committee, but not answered, was whether people actually save time or merely travel further, and if they do save work time is it transformed into productive use no matter how small that saving. The Committee also expressed concern over the valuation of accident savings and suggested that other factors such as land take, delays during construction, and more robust testing of the capital cost assumptions (often underestimated) should be explicitly included in the evaluation.

Comment was also made on environmental and regional development factors in evaluation, as both of these, although acknowledged as being important, were placed second to road-user benefits. It was accepted that evaluation of the environment was difficult but methods were available to give rankings to these factors (for example the Planning Balance Sheet; Lichfield, 1970). Increased consultation and public presentation would help in explaining options and the implications of each. The recommendations of the Leitch Committee went some way to meeting these criticisms (Department of Transport, 1979*a*).

In their proposed Comprehensive Framework for the Appraisal of Trunk Road Schemes, the justification for the scheme should include the identification of individual groups affected, a comprehensive and flexible approach that balanced the worth of a project against the extent to which each impact meets a series of standards, regulations and procedures, as is the case in the USA. The assessment should be comprehensible to the public and command their respect. The balance of assessment was switched from its overriding economic base towards a comprehensive framework which relied more on judgement, with full consultation at all stages and included other factors such as the environment in decision-making.

The Leitch Framework shared many common features with the broader based evaluation procedures developed in planning. In this sense, another strong link was being proposed between transport and planning evaluation. Most similarity was apparent with Lichfield's Planning Balance Sheet Appraisal (Lichfield, 1970; Lichfield *et al.*, 1976) which had been extensively used with the Roskill Commission's Inquiry. Essentially, the PBSA is a modified version of social cost benefit analysis (SCBA) as it deals with monetary valuations where possible, but it differs in that it explicitly differentiates between the various incident groups who gain or lose from a particular investment decision. As such it responds to two problems apparent in SCBA. First, SCBA does not attempt to measure the redistribution of resources and the equity implications of the transactions between different sectors in society. In private investment evaluation and project selection this may not be important, but where broader planning and societal goals are at stake, then it becomes a political consideration. Secondly, although monetary values are used wherever possible, intangibles are assessed without allocating financial values to them. Aggregation takes place where possible, but the Planning Balance Sheet is presented in its entirety. This presentation is important as it allows the decision-maker to see which factors are the most important and some assumptions become more transparent. However, it gives no guidance to how these factors might be incorporated into a decision, other than that impacts on groups might be weighted to account for the equity considerations.

In Hills' Goals Achievement Matrix (Hills, 1973), the evaluation moves from economic measures to a statement of achievable objectives that have been explicitly set prior to the assessment. Each goal is specified and weighted according to its importance, and these goals are set against those sections of the community affected by the consequences of each proposal being assessed. Community incidence groups can be identified by income, social criteria or location, but again each group is weighted for each objective individually or for all objectives combined. Evaluation is by monetary means where possible or by other units, and aggregation is used if the measurement units are compatible.

As with the PBSA and with the Leitch Comprehensive Framework for Appraisal, the whole matrix is presented to the decision-maker as well as the overall interpretation of the evaluation. GAM takes evaluation away from its economic base in the rationality of neo-classical economics and places it firmly in the political arena, where different objectives can be valued according to their importance and where the impacts of decisions on particular incidence groups can be calculated.

COMMENT ON EVALUATION IN THE 1970S

These three matrix approaches to evaluation acknowledge the wide range of different impacts in public sector decisions and the fact that many of those impacts are inherently non-comparable. The intention is that the decision-maker can choose a best decision judgementally after reviewing the spectrum of different impacts and making the necessary trade-offs between them. More fundamentally, the Leitch Committee did not address the possible conflicts between the role of the analyst and the decision-maker in public policy. The forecasting process and the analysis of traffic including economic evaluation had all played an important role in demonstrating the demand for new roads. This analysis had been instrumental in gaining Treasury approval and Parliamentary support for the trunk road programme.

However, in practice, as Starkie (1982) comments, 'the combination of long gestation periods for highway schemes and the time taken to develop a standard evaluation procedure meant that economics had had little influence on the shape or sequencing of the trunk road network *constructed* by the early seventies.' The actual influence of COBA and matrix appraisal frameworks on the 2,000 miles of motorway and trunk roads built before the mid-1970s was minimal. Starkie (1982) suggests that it is also debatable whether COBA, even when it was fully operational, has had much impact on the shape and size of the more recent roads programme – 'The decision to invest in a road scheme was at the end of the day a matter of judgement alone.' There seems to be a kind of inevitability about the outcomes of various decisions, and analysis may only be appropriate where it supports the preferred outcome.

Perhaps this dilemma is best illustrated with reference to the decision of the Roskill Commission. The first mistake made was the exclusion of Stansted from the shortlisted sites on the grounds that it was inferior to Nuthampstead. Stansted had been the favoured location of the London airports lobby (consisting of the airlines, the Ministry of Aviation, later the Civil Aviation Authority, and the British Airports Authority) for the previous twenty years, and it was this group that was instrumental in obtaining the Maplin Review. The second error was to allow Buchanan to submit a minority view in favour of Maplin when the rest of the Commission supported Cublington. Any recommendation would have had more weight if it was supported by all the members of the Commission. Also of importance was Buchanan's rejection of the basic economic arguments used by the Commission in favour of a broader planning-based judgement.

The lessons from both the Roskill Commission's recommendation and from the broader use of economic evaluation in transport decisions are clear. To implement successfully any recommendation, it must be unequivocally supported by all members of the Commission or Committee and it must also have strong political support from Westminster. Any weakness in that united front will allow other interest groups to intervene, and to delay any decision being made. Delay is often equivalent to the abandonment of a decision. Intervention can either come from interest groups within the sector, as with the London airports lobby, or from the general public, as with the protests over the roads programme.

With respect to major or controversial investment decisions in the transport sector, action will only take place if all the relevant parties and the government are in agreement. The strength of the case also needs to coincide with the appropriate timing for action, as unpopular decisions are not likely to be taken immediately prior to a national or local election. The opportunities for action are therefore limited.

## Evaluation and the Environment in the 1980s

The level of debate in the 1980s has generally been at a much more subdued level than in the previous two decades, when the cost benefit analysis procedures were being set up, when there were a series of large scale public inquiries, when the protest movement was most active, and when the procedures were reviewed and modified. The 1980s also marked the end of the first generation of motorway construction which had started with the Preston bypass in 1958 and ended with the completion of the M25 around London in October 1986. Certain changes have, however, taken place which epitomize three of the major concerns of the last 10 years.

One feature of all decisions has been the excessive time taken from project genesis to completion, and the balance between the time taken for preparation (too short) and the time for gestation (too long). On average, Department of Transport schemes were taking 15 years, whilst local authority schemes took seven years (House of Commons, 1990). In both cases the actual construction time accounted for two years. Public discussions took about 30 months, with planning permission, inquiry and compulsory purchases taking a further three years. Delays were taking place in Department of Transport procedures between consultations and the publication of the preferred route, and between the close of inquiry and the putting out of contracts. In their evidence to the Transport Committee (House of Commons, 1990), the Department of Transport claimed that they could reduce the total time taken by four years with a halving of the time from entry to programme and public consultation (from six to three years), and between consultation and publication of the preferred route (from two years to one year). This move towards greater internal efficiency seems a more appropriate means to streamline procedures than the earlier attempts to remove some of the public consultation stages, so that argument could only take place on mitigating circumstances and compensation (Institute of Civil Engineers, 1983). The public inquiry itself only accounted for about a third of all the time, even in the largest schemes. Other proposals (NEDO, 1984) included the use of investigating magistrates (as in France) rather than inspectors, where the rather passive role of the inspector in receiving representations and synthesizing masses of information would be replaced by a more active investigative role at the inquiry. This proposal was not acted upon.

The reduction in time taken for the inquiry process also reduces the extent of blight caused by road proposals to people living in corridors where the road might be located (the safeguarded routes). The compensation rules are very strict with market price being paid for properties that are subject to a compulsory purchase order. Again the experience from France suggests that a premium of 10–20 per cent would help alleviate the disruption caused by a forced move. The compensation procedures in Britain seem to be fairly restrictive both in terms of eligibility and the levels of recompense actually paid. More efficient inquiry procedures and more generous compensation terms would help reduce the direct effects of new transport infrastructure on individuals directly or indirectly affected.

A second issue has been the increased use of Parliamentary Procedures to speed up the planning process. It is agreed that major developments involving public resources should be assessed and justified in public to determine whether the project is needed, to assess the opportunity costs and the environmental impact, and to compare the selected alternative with other possibilities. But is the same process required when private resources are being used, where the risk is to the developer, where confidence has to be maintained by investors and where long lead times would reduce the interest of the private sector in the project? The Select Committee procedure has been used for approving major transport investment projects such as the Channel Tunnel and the new M25 Bridge which duplicates the Dartford Tunnels. The Commons and Lords Select Committees operate to a very strict timetable and have the responsibility for allocating construction and operating rights as well as land acquisition powers to the developer. Public hearings are only available for those directly affected, and there is no discussion of the wider impacts of the scheme, nor is there a requirement for the major parties to present their case. There is no debate over the alternatives, only the actual scheme, and even then it is not a question of whether it will be built but of minor changes to the proposal. The final decision lies with Parliament.

The Channel Tunnel Bill was a Hybrid Bill sponsored by the Department of Transport, Eurotunnel, British Rail and Kent County Council. The Commons Select Committee had seven members and its remit was to resolve the problems raised by the groups submitting evidence, but not to challenge the basic proposal in the Bill. In the House of Lords there was only one meeting of the Select Committee which concentrated on the most important problems identified by the Commons Select Committee. The whole review process took under one year and the Bill received the Royal Assent in July 1987, and the Channel Tunnel should open for public service in 1994.

The third issue has been the environmental effects of proposed road schemes and their assessment. The Standing Advisory Committee on Trunk Road Assessment has turned its attention to urban road appraisal, and the measurement and evaluation of the environment (Department of Transport, 1986a). Their recommendations were that road appraisal in urban areas should be integrated with the wider land-use and environmental impacts, and that consultation should take place at an early stage with the public and authorities directly affected in the proposals. Each project should have a clear statement of objectives together with the relevant transport problems and a range of feasible options which would satisfy those objectives. Appraisal should cover traffic, economic evaluation and an assessment of environmental and social impacts. The basic framework proposed by Leitch (Department of Transport, 1977b) had been extended to include a wider range of impacts together with an explicit declaration on the objectives of each scheme, which in turn reflected a move towards Goals Achievement Matrix analysis (see page 60). Evaluation had moved away from single-criteria analysis to multi-criteria analysis. Although the government's response was not fully supportive, many of the recommendations have been accepted (Department of Transport, 1986b).

Since the report of the Jefferson Committee to the Departments of the Environment and Transport (Department of Transport, 1976*b*), a restricted interpretation of the environment has been included in the assessment. Environmental costs have been internalized within the decision-making process and only local factors have been considered. The broader effects of transport on pollution, the use of resources, and development are largely ignored (table 3.6). The more recent *Manual of Environmental Appraisal* (Department of Transport, 1983) develops a list of environmental factors and adds to it assessment of heritage and conservation areas as well as ecological factors.

In June 1985, the EC issued the Directive on the assessment of the effects of certain public and private projects on the environment, including the construction of motorways and express roads – interpreted in the UK as trunk roads over 10 km in length (Article 4(1)) – and other roads or urban development projects at the discretion of individual Member countries (Article 4(2)). The Directive took effect from July 1988 (EC Council Directive, 1985). The SACTRA Report (Department of Transport, 1986*a*) concluded that their recommendations are compatible with the requirements of the Directive. However, the Directive is explicit in specifying a much wider range of factors that should be included in any Environmental Impact Assessment. Article 3 requires an assessment of the direct and indirect effects of a project covering the following factors:

- human beings, fauna and flora;

- soil, water, air, climate and landscape;

- the interaction between the factors mentioned above;

- material assets and the cultural heritage.

Article 5 requires the developer to supply information on the following topics and to have regard to current knowledge and methods of assessment:

- a description of the project comprising information on the site, design and size of the project;

- a description of the measures envisaged in order to avoid, reduce and, if possible, remedy significant adverse effects;

- the data required to identify and assess the main effects which the project is likely to have on the environment;

- a non-technical summary of the above information.

The EC Directive provides clear guidance on how a comprehensive Environmental Impact Assessment (EIA) can be prepared. EIA refers to the evaluation of the environmental improvement likely to arise from a major project significantly affecting the environment. Consultation and participation are integral to this evaluation, the results of which must be available in

TABLE 3.6. Environmental factors and transport in UK.

| | Importance of Transport | Effects |
|---|---|---|
| **1. POLLUTANTS** | | |
| Carbon dioxide | 16% | Global warming |
| Nitrogen oxides | 45% | Acid rain |
| Sulphur dioxide | 5% | Acid rain, bronchitis |
| Carbon monoxide | 85% | Morbidity, fertility |
| Benzine | 80% | Carcinogenic |
| Lead | 50% | Mental development |
| Hydrocarbons | 28% | Toxic trace substances |
| **2. RESOURCES** | Consumption | |
| Oil | 47% | Depletion of natural resources |
| Land Take | | 6 ha of land per km of motorway |
| Ecology | | Landscape and SSIs destroyed |
| Ecosystems | | Water quality, flood hazards, river systems modified |
| Accidents | 5,000 deaths 63,000 serious injuries | Pain, suffering, grief |
| **3. ENVIRONMENT** | | |
| Noise | | Stress, concentration, health |
| Vibration | | Historic buildings |
| Severance | | Dividing communities |
| Visual impact and aesthetics | | Changes in physical appearance |
| Conservation and townscape | | Preservation |
| **4. DEVELOPMENT** | | |
| Regional development | | Location of industry |
| Local economic impacts | | Income levels, employment, social impact |
| Congestion | | Delay, use of resources |
| Urban sprawl | | Traffic generation, induced development |
| Construction effects | | Blight, property prices, compensation |

the form of an environmental impact statement (EIS), before a decision is given on whether or not a project should proceed (Wood and Jones, 1991).

The *Manual of Environmental Appraisal* meets the requirements of the EC Directive and public inquiries are held for all major schemes to which there are objections, and the environmental assessment will be considered at that inquiry. There is an enormous literature on EIAs in the USA where there has been a requirement since 1969 (National Environmental Policy Act) for Federal agencies and those in receipt of Federal funds to prepare Environmental Impact Statements (EIS) for 'major actions'. Such actions include any alterations in land use, planned growth, travel patterns or natural or man made resources. All new freeways, expressways or belt highways, or reconstructed arterial highways providing substantially improved access to an area are included within the NEPA requirements.

In the UK, the Town and Country Planning regulations (July 1988) make it compulsory for some types of project to have an EIA to accompany the planning application. This requirement is in line with the EC Directive and ensures the EIA's compatibility with existing planning procedures. Schedule One includes major transport infrastructure proposals and aerodromes with runways over 2100 metres as projects requiring an EIA. If the proposed development is of sufficient size or is in an environmentally sensitive area the local authority may require an Environmental Impact Statement (EIS) – a summary document. Once the EIS has been submitted, the normal planning procedures are followed with the exception that the local authority has 16 weeks to make a decision rather than the normal eight weeks. The recent report of SACTRA on the environmental effects of proposed road schemes and their assessment (Department of Transport, 1992*b*) comments on the full implementation of the EC Directive and on the possibilities for the valuation of environmental factors in assessment.

As the requirement for a full EIA is now part of EC policy and as it has been included within the planning process, it would seem logical to extend explicitly the requirement to the assessment of all new road schemes in the UK. EIA allows the broader environmental issues to be fully presented without lessening the importance of economic factors in assessment. A full EIA is compulsory in France and it also forms an integral part of the road planning process in West Germany.

There seems to be a strong case for the extension of existing economic-based evaluation methods to encompass a broader range of environmental, social and distributional factors. The SACTRA Report goes some way towards arguing the case for such a change (Wood, 1992), but stops short of a full commitment because of many unresolved issues, such as global and local impacts, the complexity of the environment, valuation of the environment and the general unreliability of forecasts. Yet decisions still have to be made and many controversial issues (for example Twyford Down near Winchester and Oxleas Wood in London) have been resolved without a full environmental appraisal. This in turn has resulted in confrontation between European and national government, and may provide environmental groups with a new impetus in the 1990s.

# Chapter 4

# Contemporary
# Transport Planning

## The Market Alternative to Transport Provision

Since the 1960s governments have always played a major interventionist role
in transport decisions. The underlying philosophy was that transport should
be made available to meet the needs of the population and that everyone could
expect a minimum level of mobility, almost as a right. To this end, once the
basic motorway and trunk road networks had been established, investment
was switched to public transport in the form of capital and revenue expenditure. In the last decade the role of government has been significantly reduced
and market forces have been allowed to determine both the quantity and to a
great extent the quality of public transport services. Government policy on
transport maintains that services should wherever possible be provided by the
private sector, services should be determined competitively, not in a coordinated fashion, and fares should be market priced. Coupled with these changes
is a move towards greater precision in defining the objectives of public
transport enterprises, particularly financial performance objectives and quality of service standards. In practice the reduced role of government is more
apparent than real as it could be argued that central government has become
more powerful.

It is at the local level that the impacts have really been felt, with the abolition
of the Greater London Council and the Metropolitan County Councils, the
protected expenditure levels, and the deregulation legislation. Similarly, where
intervention has taken place, it has been targeted towards individual initiatives
(for example Urban Development Corporations) to correct perceived market
distortions.

In transport terms the 1980s was the decade of the motorist with the costs
of driving being significantly reduced and the growth in company financing of
motoring becoming one of the major private subsidies. In addition to the

phenomenal growth in car traffic of some 40 per cent over the decade, there was a parallel increase in overseas travel, principally by air on business and holidays. Over 25 per cent of the population take two or more holidays, and expenditure on leisure now accounts for 17 per cent of all household expenditure. The numbers of overseas holidays taken by UK residents trebled from 7 million to 21 million in the period 1976–1988. Society is now in transition from one based on work and industry to one in which leisure pursuits dominate – the post-industrial society. This society will be highly mobile and depend increasingly on the car and technology.

The decade was one of optimism, at least until the events of 'Black Monday' in October 1987, with unprecedented growth in house prices, incomes and quality of life for those in employment and not on fixed incomes. It is difficult to conclude whether this wealth was policy induced or whether it was a longer-term cycle of growth in the world economy. Nevertheless, it was a decade of almost continuous activity which has brought about the most fundamental changes in transport policy seen in Britain this century.

- The planning framework within which transport policy has been structured has been dismantled with the intermediate tier of local government being abolished. The government now deals directly with the districts (and Boroughs in London) and the statutory transport undertakings (for example British Rail and London Transport).

- The provision of bus services outside London is now fully deregulated with services being provided wherever possible by the market at the appropriate price. Subsidy is specific to routes where need has been identified and where services cannot be provided by the market. In London competitive regulation still exists but full deregulation is likely in 1994 (Banister, 1992b).

- Transport enterprises have been sold to employees, to management, to other companies and to the general public. Most transport related enterprises have been denationalized.

- A series of measures have been introduced to improve safety, such as compulsory seat belt wearing, changes in the motorcycle test, and changes in the drink-drive regulations.

- Traffic in towns and congestion have been identified as key problems and some remedial measures have been introduced, such as wheel clamps and the towing away of illegally parked vehicles, 'Red Routes' in London, and increases in parking fines with enforcement being carried out locally by the district authority or private contractors.

- The environment is now seen as the next major political issue, but direct action in the transport sector has been restricted to a reduction in tax levels on unleaded petrol.

The main actions of the Conservative governments of the 1980s are summarized in table 4.1. Underlying all these changes has been one consistent objective, namely to reduce the levels of public expenditure in transport, particularly on the revenue side. Although many of the policies have been justified on other criteria, such as ideology (for example deregulation), democracy

TABLE 4.1 The Thatcher years, 1979–1991.

| *1979–83* | *May: Conservatives elected majority 43* |
|---|---|
| Falklands War 1982 | Crude oil $18–24 per barrel<br>Increase in public transport fares 15–25%<br>Fares Fair campaign in London and other cities<br>Armitage Report on Lorries, People and the Environment<br>Transport Act 1980 – Deregulation Part I<br>Transport Act 1981<br>Transport Act 1982<br>Collapse of Laker 1982<br>Seat belt wearing mandatory – January 1983<br>Wheel clamps and travel cards in London – May 1983 |
| *1983–87* | *June: Conservatives re-elected* |
| Miners Strike March 1984<br>National Docks Strike<br>End of Miners Strike March 1985 | Transport Act 1983 – Protected Expenditure Levels<br>London Regional Transport Act 1984<br>Local Government Act 1985 – Abolition of GLC and MCCs<br>Capital cards in London – January 1985<br>Transport Act 1985 – Deregulation Part II<br>Anglo-French Channel fixed link – February 1986<br>Terminal 4 opened at Heathrow and Piccadilly extension<br>Crude oil $10 per barrel<br>Completion of M25 – October 1986<br>British Airways privatized – January 1987<br>Rolls Royce privatized – May 1987<br>British Airports Authority privatized – June 1987 |
| *1987–91* | *June: Conservatives re-elected* |
| Collapse of Stock Exchange and Gales October 1987<br><br><br><br><br><br><br><br>Iraqi Invasion of Kuwait August 1990 | Docklands Light Railway opened – July 1987<br>City Airport opened – October 1987<br>Merger BA/British Caledonian – November 1987<br>National Bus Company privatization – April 1988<br>Central London Rail Study – January 1989<br>Roads Programme expanded – May 1989<br>New Roads by New Means – May 1989<br>Autoguide Electronic Route Guidance piloted – July 1989<br>London Assessment Study proposals axed – March 1990<br>Crude oil rises to $26 per barrel<br>Roads for the Future – February 1990<br>New terminal at Stansted Airport – Spring 1991 |

*Source*: Banister (1990).

(for example widening share ownership), efficiency (for example private enterprises are more cost effective than public ones), and accountability (for example to shareholders), it is the Treasury which has been the prime mover behind the radical transport policies.

Even capital expenditure on renewal and new transport infrastructure has been severely constrained by Treasury imposed limits. Over the last ten years replacement expenditure has been delayed, and this in turn has led to increasing unreliability in the system. As can be seen in table 4.2(*a*), capital expenditure on transport infrastructure remained stable in real terms to 1987, despite increases from 1982–1984. It was only in the period 1988–1991 that a substantial increase took place (+21 per cent).

The low levels of capital expenditure in the mid-1980s coincided with record growth in the demand for travel by public and private transport, with the annual growth rate in the numbers of new cars peaking at 6 per cent in 1989. The second feature of the changing pattern of capital expenditure has been the switch from public transport investment to investment in roads. In 1981 public transport accounted for 16 per cent of total capital expenditure, but by 1991 this figure had declined to 5 per cent; in real terms the 1991 figure is about 37 per cent of the 1981 figure. The figures (table 4.2(*a*)) reflect the recent relaxation of public expenditure constraints and new capital schemes for roads, railways and London Transport (Department of Transport, 1989*b*; House of Commons, 1990), but much of the present increase in capital expenditure could be seen as merely catching up on previous under-investment in capital renewal.

Increasingly the government is looking to the private sector for capital. The scope for the private sector in urban areas may be limited as construction costs are high, development lead times are lengthy, risks are high, and returns are likely to be low unless the investment puts the developer in an effective monopoly position. Development rights along the proposed route may make private investment more attractive as would joint ventures with the government underwriting the risk (Department of Transport, 1989*c*).

It is on the current expenditure side that the government has been most active. In the early 1980s there was a concern over the increasing levels of public support for both bus and rail services with subsidy levels for each amounting to about £1,000 million. The market philosophy is to ensure that all services are provided at a level and at a price which is determined competitively. Intervention should only take place when the market is seen to fail, for example for social need, but even then the intervention should be specific and clearly identified. This means that subsidy levels should be minimized and that there should be no cross subsidization between services. Protected expenditure levels were set for each conurbation and these were effectively the maximum permitted subsidy levels. A balance had to be established between the interests of travellers and those ratepayers/polltax payers and taxpayers who contribute to the subsidy. The net result has been significant increases in fares on all bus

TABLE 4.2(*a*). Levels of public capital expenditure on transport infrastructure, Great Britain (£ million).

| Year | National Roads | Local Roads | Public Transport | Ports | Airports | Deflator | Real 1985 Prices |
|------|------|------|------|------|------|------|------|
| 1981 | 720 | 541 | 257 | 16 | 36 | 1.253 | 1967 |
| 1982 | 874 | 661 | 260 | 14 | 45 | 1.164 | 2168 |
| 1983 | 864 | 703 | 284 | 13 | 42 | 1.106 | 2108 |
| 1984 | 908 | 730 | 318 | 13 | 30 | 1.057 | 2113 |
| 1985 | 920 | 837 | 118 | 12 | 46 | 1.000 | 1933 |
| 1986 | 958 | 853 | 69 | 12 | 39 | 0.966 | 1865 |
| 1987 | 1100 | 869 | 105 | 8 | 42 | 0.921 | 1956 |
| 1988 | 1145 | 1104 | 65 | 8 | 79 | 0.863 | 2072 |
| 1989 | 1493 | 1270 | 134 | 15 | 96 | 0.806 | 2424 |
| 1990 | 2024 | 1039 | 171 | 12 | 83 | 0.759 | 2527 |
| 1991 | 2084 | 1186 | 167 | 9 | 79 | 0.712 | 2510 |

TABLE 4.2(*b*) Levels of public current expenditure on transport, Great Britain (£ million).

| Year | Revenue Support for Road Public Transport | Concessionary Fares – All Modes | Public Service Obligation for British Rail | Real 1985 Prices |
|------|------|------|------|------|
| 1981 | 395 | 193 | 800 | 1739 |
| 1982 | 510 | 241 | 831 | 1841 |
| 1983 | 523 | 260 | 855 | 1233 |
| 1984 | 555 | 274 | 827 | 1750 |
| 1985 | 465 | 290 | 832 | 1587 |
| 1986 | 406 | 296 | 737 | 1390 |
| 1987 | 332 | 324 | 775 | 1318 |
| 1988 | 337 | 350 | 473 | 1001 |
| 1989 | 297 | 380 | 501 | 949 |
| 1990 | 328 | 419 | 602 | 1021 |
| 1991 | 401 | 448 | 892 | 1240 |

and rail services, both prior to and after deregulation. Subsidy levels were reduced by about 50 per cent in the period immediately following deregulation (table 4.2(*b*)), but they have now crept back to the levels of the early 1980s (in cash terms). Ten years ago there was an approximate balance between government expenditure on capital and current accounts (table 4.2): the present figure is 2 to 1 in favour of capital expenditure on a total budget that has remained constant in real terms.

One continued area of concern for government is the increase in levels of concessionary fares (table 4.2(*b*)). At present eligibility is based on fairly broad social groups (for example the elderly, school children or disabled), but it is realized that not all individuals within each concessionary group need the same level of concession. More sensitive methods are being identified to allocate concessionary fares to those in need. The net effect would be a reduction in the total level of subsidy.

If the government is to be consistent in this policy then the subsidy to company cars should also be reduced or eliminated. Over 50 per cent of all new car registrations are in a company name and many organizations also pay for insurance, tax and petrol for their employees. Motorists who do not have to pay the true costs of their travel will use the car more frequently and make more trips. About a third of all vehicles on the road receive some form of company assistance. Tax liability was increased over the four years 1987–1990 by 200 per cent, but the benefit is still over £2,000 for a 1200 cc car and over £5,000 for a 2,000 cc car for an individual paying tax at 25 per cent (Reward Group, 1990). Apart from the inconsistencies in subsidy policy between the reduction in subsidy for public transport and the increase in subsidy for the company car, the effectiveness of any policy designed to reduce the non-essential use of the private car (for example road pricing) is diminished if the user is shielded from paying those additional costs. For the market economy to work effectively, competition in both the public and private transport sectors must be fair.

Transport policy over the last decade has been driven by the desire to reduce levels of public expenditure on transport, particularly on the revenue side, and to reduce the role of government in determining policy. Overall, the government has been successful, at least in reducing levels of current expenditure by some 29 per cent in real terms (1981–1991). Demand for travel, principally by the private car but also by rail (since 1985), has continued to increase at an unprecedented rate and road user expenditure on transport now contributes over 12 per cent of all Exchequer revenues, some £17 billion in 1989. Over the decade this figure has risen by 184 per cent in actual terms (62 per cent in real terms), and this has been one factor in the ability of the government to reduce levels of direct taxation.

Apart from the increased contribution of transport to Exchequer revenues and the decreases in public expenditure on transport, this sector has been the testbed for the three main policy innovations of the last ten years.

- to transfer industries from public sector to private sector ownership: privatization or denationalization;

- to introduce competition into public transport services and reduce levels of public subsidy: regulatory reform;

- to encourage more private capital in major transport investment projects, particularly in infrastructure.

## PRIVATIZATION

After 1945 there was an extensive programme of nationalization which formed part of Herbert Morrison's plans for the reconstruction of British industry.

Almost all of the transport enterprises were placed in public ownership. Since 1979 the Conservative government has returned most of these nationalized industries to the private sector, either in their entirety or through partial sales. The reasoning behind this policy is partly ideological and partly financial. The ideological argument is that organizations are more efficient in the private sector, with the normal pressures that competition brings. Public bureaucracies are inefficient suppliers of services because there are no measures of productive efficiency. In the private sector there is a simple measure of performance, namely profitability. The financial argument is that the Exchequer can gain through the sale of companies, and it can save through reductions in the levels of public support paid by means of subsidy or lending to service capital debts.

Giving more freedom to management is crucial as it allows capital to be raised in a variety of ways on the financial markets. Private sector enterprises are no longer constrained by the external financing limits imposed by government in the early 1980s. Deficits sustained by public transport operations have had an unwelcome effect on the government's public sector borrowing requirement. Privatization also contributes directly to exchequer revenue and reduces the public sector borrowing requirement as the ratio of government expenditure to national income is reduced. Supply side economics would also suggest that inflationary pressures are reduced, tax cuts can be introduced and this in turn would allow more private investment to take place so that productivity, output and profitability can all increase. This powerful logic is countered by the argument that these gains are short term. The longer-term impacts of selling national assets may result in higher taxation as the government is foregoing revenue from these nationalized industries. In effect the government is borrowing against future income streams (Rees, 1986).

From table 4.3 it can be seen that there have been two distinct phases of transport privatizations which have coincided approximately with the first two Thatcher administrations. During the first phase, the privatizations were small in scale often involving management buyouts (for example the National Freight Corporation), or a single buyer (for example for Sealink), or a limited stock exchange flotation (for example Associated British Ports). The second phase has been much larger in scale and reflects the strong ideological commitment to privatization. It is here that the government has, through extensive publicity, encouraged wider share ownership and has made provision for the smaller investor to 'buy' part of the newly privatized company, often at an advantageous price. The four transport privatizations in Phase I (1982–84) realized about £500 million and the four sales in Phase II (1987–88) realized nearly £4,000 million (table 4.4). Further transport privatizations include British Rail and London Transport, and legislation has been introduced in 1993 to permit franchising of selected British Rail routes, but with control of the track remaining under the management of Railtrack. It is in effect a partial privatization.

TABLE 4.3. Privatization in transport in the United Kingdom.

| | |
|---|---|
| *National Freight Corporation:*<br>February 1982 | Road haulage operator<br>Sold for £53 m to a consortium of managers, employees and company pensioners. The government paid back £47 m to the Company's pension fund to cover previous underfunding |
| *Associated British Ports:*<br>February 1983<br>April 1984 | Ports and property development<br>Part of equity sold<br>Remainder sold by tender offer<br>£34 m raised |
| *British Rail:*<br><br>July 1984 | Some non-essential assets sold<br>British Rail Hotels sold in 1983 for £30 m<br>Sealink Ferries sold to Sea Containers Ltd for £66 m. Proceeds retained by BR with subsequent adjustments to the borrowing limits |
| *Jaguar:*<br><br>July 1984 | Luxury car manufacturer which had become a subsidiary of British Leyland<br>Sold for £294 m |
| *British Airways:*<br>January 1987 | One of the leading international airlines.<br>Sold for £892 m. |
| *Rolls Royce:*<br>May 1987 | Aeroengine business bought by the government in 1971.<br>Sold for £1,360 m. |
| *British Airports Authority:*<br><br>July 1987 | Operates seven of the principal airports in Britain including Heathrow and Gatwick<br>Sold for £1,280 m |
| *National Bus Company:*<br><br><br>December 1988 | Consists of 72 subsidiaries which run most bus services outside the main metropolitan and other urban centres<br>Sale completed with gross proceeds of £323 m. Net surplus to government after all debts paid of £89 m |

*Source*: Banister (1990).

TABLE 4.4. Public expenditure in transport in real terms (£ billion).

| | *1981–82* | *1983–84* | *1985–86* | *1987–88* | *1988–89*<br>(estimate) |
|---|---|---|---|---|---|
| Transport | 6.9 | 6.4 | 6.1 | 5.7 | 5.6 |
| Privatization | −0.7 | −1.4 | −2.9 | −5.1 | −5.7 |

*Notes*:
1987–88 price levels.
Privatization proceeds are from all sectors not just transport.

*Source*: Based on HM Treasury (1989).

These are the last two major transport enterprises still in public ownership. At that time the thirty-year cycle, during which all major transport enterprises were in the public sector, will have been reversed and all 'public' transport assets will now be in the private sector (Banister, 1990).

REGULATORY REFORM

Deregulation of the bus industry has been the most important policy switch since regulation was first introduced in the 1930s to protect a vibrant growth industry against predatory practices. Deregulation as it affects all urban and non-urban areas outside London has been introduced in two stages. The 1980 Transport Act deregulated long-distance bus services (those with a minimum journey length of 48 kilometres), excluded small vehicles (less than eight seats) from licensing requirements provided that they were not being run as a business, and introduced the idea of trial areas where there would be no requirement for road service licences. The 1985 Transport Act has effectively created a trial area for the whole country except London. The commercial network only has to be registered, and the local authority has no control over the operator provided that he or she runs the services as registered. The subsidized network complements the commercial network with each service being put out to tender by the local authority who has to specify the route, levels of service and fares (in most cases). London has been given an intermediate position under the terms of the London Regional Transport Act (1984) with only limited competition through competitive regulation (Banister, 1985; Glaister, 1990).

This two-tier system treats each route on its own merits and there is no commitment to a network of services. The government argues that competition will bring greater efficiency into the public transport industry and an end to the extensive practice of cross subsidization. The main impetus to change has been the concern over the escalating costs of grants and subsidies to the bus industry, which totalled over £1,000 million in 1983.

The deregulation of bus services under both full deregulation (outside London) and competitive regulation (in London) has led to significant cost savings per vehicle kilometre, amounting to some 40 per cent in the Metropolitan Areas and 20 per cent in London (table 4.5). It should be noted that although these figures are in real terms, they only relate to the supply side. Costs in terms of pence per passenger journey have increased by 24 per cent in the Metropolitan Areas and by 13 per cent in London (Department of Transport, 1991b). These savings in operating costs include reductions in wage levels as well as improvements in productivity. Savings have been achieved through direct reductions in basic wages, lower wage levels for new staff, and lower wage rates for minibus drivers. The Local Authorities have also provided generous voluntary redundancy schemes which have enabled the

Passenger Transport Companies to trim staff levels, particularly in engineering and maintenance staff (–41 per cent), but also in platform staff (–7 per cent). In London there has been less scope for reductions in staff costs as salaries in the capital have been at higher levels historically, and there have always been problems in recruitment. In addition to the downward pressures in wages, there have been significant improvements in productivity with increases in shift working, greater time spent actually in service, and greater use of the available vehicles, in particular with the introduction of minibuses. The crucial long-term question is whether this level of savings in the costs of providing the services can be maintained. Most of the savings were made in the first three years of deregulation.

For competition to take place in the bus industry, the market must be contestable. It seems that the conditions of contestability do exist as sunk costs are low and all operators have access to the same technology (Banister, Berechman and De Rus, 1992). The market is less contestable when barriers to entry are considered, such as access to bus stations, the availability of service information and experience, and the requirement for 42 days notice for changes in service. These factors have made it difficult for 'hit and run' operations where the new entrant sees a market opportunity and enters the market for a short while to make a profit and then withdraws when the incumbent operator reacts. The market position of small operators seems to be with the tendered services and not on the commercial network. Competition on the main corridors of the commercial network in the Metropolitan Areas appears to be restricted to the larger operators where competition takes place among equals and where the incumbent advantages are limited. This means that smaller operators may have a greater potential to compete in London where services are contracted out. The smallest group of bus operators (each covering under 5 million kilometres on local services per annum) increased their share of local markets from 13 to 17 per cent between 1985 and 1987. Since then, their overall share has remained fairly static, though their importance has increased further in London and the Metropolitan Areas, principally through successfully competing for the subsidized network (Department of Transport, 1991*b*).

The natural sequence in the competitive market may be the re-establishment of oligopolistic or monopolistic operations, and it will only be on the tendered services that the small operators will be able to compete as the successful applicant is granted limited monopoly rights through the bidding process. In the Metropolitan Areas there may be a few large operators running the vast majority of services with small operators running special niche services and some tendered services. It may be in London that the greater variety of operators will be found as all the competition takes place off the road, and small operators can compete with the large operators on equal terms through the tendering process.

TABLE 4.5. Summary of changes over the first five years of deregulation in London and the Metropolitan Counties.

|  | *Full Deregulation: Metropolitan Counties* | *Competitive Regulation: London* |
| --- | --- | --- |
| Passenger journeys | −26.2% | +3.9% |
| Vehicle kilometres | +12.9% | +11.4% |
| Fares[1] | +31.8% | +12.0% |
| Proportion tendered | 14% | 40% |
| Stability | Improving | High |
| Staff employed | −26.0% | −23.0% |
| Passenger receipts[1] | −12.2% | −9.1% |
| Revenue support[1] | −61.3% | −18.0% |
| Concessionary fares[1] | +19.3% | +8.3% |
| Operating costs[1] | −40% | −20% |

*Note*: 1.   Changes denoted in real terms from 1985 to 1991.
The Metropolitan Counties include all major British cities – West Midlands, Greater Manchester, Merseyside, West Yorkshire, South Yorkshire, Tyne and Wear, and Strathclyde.

*Source*: Department of Transport (1991*b*).

The Metropolitan Areas have been the crucial test of the 1985 Transport Act and the main effects seem to have been a reduction in revenue support on buses at no great loss to service frequency (table 4.5). Tyson (1988, 1989, 1990) estimates that about 70 per cent of this saving to the tax and rate payer has been passed onto the traveller through higher fares. However, due to the uncertainty caused by a fairly traumatic period of transition, there has been a significant downturn in patronage which has again had implications on the profitability of bus companies. Permanent damage may have been caused to the quality of bus services.

The net effect of deregulation in the Metropolitan areas has been significant increases in fares (+ 31.8 per cent in real terms between 1985 and 1991) and a decline of 26.2 per cent in passenger journeys. Undoubtedly some radical change was necessary, but competitive regulation or franchising as practised in London may have been more appropriate than full deregulation, as patronage there has been increased (+3.9 per cent) and fares have risen in real terms by 12 per cent (table 4.5 and Banister and Pickup, 1990). Cost savings can be made through the tendering process and the quality of the service can be maintained through the terms of the contract.

Although significant savings have been obtained through reductions in revenue support, there have been compensating increases in the levels of concessionary fares. Over the five years of deregulation (1986–91), revenue support for bus services outside London has decreased by 53 per cent and concessionary fares have increased by 10 per cent in real terms. Despite this, the net overall saving is still considerable at nearly 30 per cent (£217 million).

If competitive regulation had been introduced in the Metropolitan Areas then the cost savings achieved in London through the increases in productivity could have been further augmented by the savings made in labour costs due to the differences in the local labour market conditions outside London. Net savings seem to have been similar under the two types of competitive regimes but the costs of full deregulation are much higher as there is much greater instability and loss of patronage. Although there seems to be a strong case to delay the decision to deregulate fully bus services in London until the longer-term effects of the two types of competitive regimes have been evaluated, the new Conservative government is likely to introduce full deregulation into London in 1994. The British experiment has been keenly observed by other European countries but it seems unlikely that they will adopt the same model of full deregulation.

## INFRASTRUCTURE

Britain completed its first generation of motorway construction with the opening of the last section of the M25 around London in October 1986. Progress has been steady since the building of the Preston Bypass in 1958, and there are now some 2,800 km of motorways. However, it seems that the 1990s may signal a significant period of new investment, not just in the motorway network. Much of the urban infrastructure is Victorian in construction and needs extensive renovation. In the transport sector several large new projects have been planned or approved – the Channel Tunnel, London's Third Airport, the electrification of the East Coast main railway and the widening of several of the first generation of intercity motorways.

Construction of new infrastructure has raised two important dilemmas. The first is the contradiction between the desire for greater mobility and meeting the ever increasing demand for travel, and the concern over the environment. Apart from needing land for new roads (6 ha of land required for each kilometre of motorway), there are also development pressures, conflicts between conservation and landscape objectives and employment and economic objectives, and a wide range of pollution and resource implications, as well as the direct impact of noise, vibration and severance. The relevant factors are complex, and although considerable debate has taken place over environmental concerns, they often only appear as an addition to the evaluation process and are rarely seen as being influential in decision-making. The high-mobility, car-oriented society is incompatible with reductions in levels of pollution, lower consumption of energy and protection of the environment.

The second dilemma concerns the financing of major new infrastructure investments. In the past, infrastructure has been seen as a public good and funded from the Exchequer. With the reductions in public expenditure, the private sector has been seen as taking more of the risk, putting together the finance, and receiving payment in the form of tolls for the use of the facility. Implementation can be expedited through the use of Parliamentary procedures which avoid the necessity for large, costly and lengthy public inquiries. Examples include the construction of a second bridge across the Severn and the new bridge to supplement the two Dartford Tunnels on the M25. The Dartford River Crossing Company (Trafalgar House, Kleinwort Benson, Bank of America and Prudential Assurance) has built a suspension bridge to provide four new lanes for southbound traffic on the M25. As part of the deal they have also taken over both existing tunnels which are now used for northbound traffic and receive all the toll revenue. The new bridge opened in 1991. The problems of finance and public inquiry did not arise and the government took no risk. Although there are controls on the levels of tolls which can be charged, the company has a maximum of 20 years to recoup costs and make a return on investment. As there is no other river crossing within 20 kilometres this return is almost guaranteed.

The most prestigious project is the Channel Tunnel which epitomizes the government's new optimism concerning the future direction of large-scale infrastructure projects, and the use of private capital. The Channel Tunnel Group-France Manche (CTG-FM) receives no public funds or financial guarantees, and so takes the risk itself. The governments have given assurances to investors that there will be no political interference or cancellation, and that the promoter has full commercial freedom to determine policy including fares levels. The Channel Link Treaty was signed in February 1986 and a Private Bill was introduced in Parliament. No public inquiry has been held, only Select Committee hearings where a tight schedule has allowed only limited discussion of many local concerns, in particular about the site of the tunnel entrance and facilities at Cheriton just north of Folkestone. Construction started on both sides of the Channel in 1987 and the tunnel should open for public service in 1994. Capital has been raised in the City and through a public share flotation, and cost estimates have doubled to nearly £8 billion. The scheme will be completed a year later than originally scheduled. It has yet to be seen whether it will make a return for its investors.

The Channel Tunnel provides the procedural model for the future. Consultation procedures, which have often delayed road proposals for up to ten years and the decision on London's third airport for nearly twenty years, have now been short circuited through the use of Parliamentary procedures. The government now takes only limited risks as the capital has been raised privately. The general public may feel that they have been ignored as they have no direct input to decisions that often affect them directly. Private companies are looking for investment opportunities which can guarantee a return or minimize any risk. Projects initiated in this way are unlikely to fit into any overall strategy and the procedure does not reduce the need for public expenditure on infrastructure. Similarly, private sector projects are very much more dependent upon market conditions and assumptions of continued growth in the demand for travel. Consequently, private sector investment should be seen as an addition to the public sector programme, not as a replacement for it.

Through radical changes in policy, the government has succeeded in reducing the levels of public expenditure on transport (table 4.4), and this has been offset by proceeds from the privatizations. The growth in demand for travel and the increase in car ownership have also resulted in record revenues for government from the transport sector.

However, even after 10 years of unprecedented activity in the transport sector there has been no statement of transport policy, except that the tenets of a market approach, efficiency and value for money seem dominant; this dominance is implicit rather than explicit. The question is whether that 'no policy' position is tenable and whether it can be maintained over the next decade.

## The Lessons for the 1990s

It can be argued that successive governments have avoided a comprehensive statement on transport policy. The Labour government attempted such a commitment in 1976 with *Transport Policy – A Consultation Document* (Department of Transport, 1976*a*), but their efforts were not implemented as there was a change of government shortly after the Transport Act 1978. The energy crisis, world recession, high inflation and unemployment all proved to be more important issues for government action.

Although the 1980s have brought a period of unprecedented activity in transport legislation, there has still been no coherent statement of transport policy. Nevertheless, the 1980s has been the decade of the car, with traffic increasing by 40 per cent and the numbers of cars and taxis by about 30 per cent. In real terms the costs of motoring have never been cheaper, and some two-thirds of UK households now have a car. To the user the car offers real advantages which alternative forms of transport can never match except in congested urban areas and over long distances. The car has unique flexibility in that it is always available, it offers door to door transport, and it effectively acts as a detachable extension to the home. This freedom is entirely consistent with the emergence of the ideologies of the new right in the 1980s.

After a decade of consistent growth in demand for transport, certain major weaknesses and inconsistencies in the policy approaches of the 1980s can be identified. Five major concerns will be covered here, and each picks up one of the major switches in policy over the last decade or one of the major issues likely to face analysts and decision-makers in the 1990s.

1. *Infrastructure investment and replacement* is required, but the practice of relying on the public sector alone and the inclusion of public expenditure on

transport infrastructure in the Public Spending Borrowing Requirement all need thorough debate. The appropriate roles of the private and public sectors in transport infrastructure provision need to be established.

2. *Congestion* is often seen as the most important problem, but this has been true over the last 20 years as well. Does the new realism actually bring any new thinking?

3. *Forecasts* have always been difficult to make as this anticipates the demand for travel either in a global sense or for particular routes where investment decisions are being made.

4. *Regulatory reform in the bus industry* has brought many short-term benefits, but at a considerable cost in terms of lost patronage. It is unclear whether competition and contestable market conditions are appropriate within transport, or whether the market is too small and natural monopolies will result.

5. *Environmental issues* have placed governments under increasing international pressure to make decisions which are environmentally benign, but these decisions are often at variance with economic efficiency and consumer choice. Can this conflict be resolved so that the environmental and economic factors are working in the same direction?

The private sector cannot replace the public sector for capital investment in the *infrastructure*, and there must be some form of partnership. The reponses to the government's Green Paper, *New Roads by New Means* (Department of Transport, 1989c) were very clear on this. The only projects which the private sector was interested in undertaking were those that might give effective monopoly control (for example the Dartford Tunnels and Bridge) or those which were necessary to attract tenants to major office developments (for example the Canary Wharf development funding the Docklands Light Railway extension to Bank). A third possibility might be where development rights would be given as part of the proposed transport investment. An example might be to allow development at a road intersection to be undertaken by the company which had paid for the construction of that road. The reasons for the lack of interest from the private sector even when there was economic growth are clear. There is always the problem of 'free riders'. Why pay for transport infrastructure when it is available at zero cost already?

The case of the Jubilee line extension from Charing Cross through the Surrey Docks and the Isle of Dogs to Stratford illustrates the problems. All these areas would have benefited from the railway and the rent levels for businesses and private landlords would have been considerably enhanced. Yet it was impossible to set up a package where all beneficiaries would have contributed to part of the construction costs. The largest developer in Docklands (Olympia and York) offered to pay £200 million, but even this contribution would be phased so that most was paid after the line was opened. Debate and dispute led to delay and may have been responsible, at least in part, for the demise of Olympia and York's prime development at Canary Wharf.

The private sector is interested in low risk projects which have a short payback period and provide a good return for their investors. Most transport projects have long pay back periods (20–25 years) with low rates of return. They also involve large capital flows early in the investment, often result in considerable capital overspend, and are expensive to withdraw from once a decision to go ahead has been taken. The history of the Channel Tunnel project also illustrates all the problems of major transport investments (Holliday *et al.*, 1991).

The government has now recognized the major role that the public sector should play, and investment in road and rail schemes has been sharply increased (1989–1991). However, much of this increase will be used to catch up with underinvestment in the early part of the decade (table 4.2), and part of it allows the local authorities to raise the money themselves through credit approvals. A second generation of road construction and upgrading is planned, including 160 bypass schemes, the upgrading of the A1 to motorway standard, and the construction of a fourth lane on most of the M25. The National Roads Plan's current investment levels are some 60 per cent higher in real terms than in 1979 (Department of Transport, 1989*b*) and the total roads budget for the period 1989–1992 is 22 per cent higher in real terms than over the last three years (an additional £4 billion).

*Congestion* is seen as the major problem facing transport planning and this increased level of investment may help alleviate that congestion. However, it seems that in the longer term additional capacity may in turn result in more demand, with the levels of cars owned and cars used per kilometre of road increasing; and road congestion will get even worse over the next decade. This problem may be insoluble, but it does raise a series of important questions and it has acted to focus attention from all interested parties.

At the planning level it strongly suggests that intervention is required. Despite the heroic efforts of traffic management schemes and more recent technological developments, the transport system is operating close to capacity for considerable parts of the day. The choice for the government is to decide whether a strategic view conflicts with a free market. Although business has clear commercial objectives concerning its own profitability, it may prefer to operate within a publicly planned environment with a longer-term horizon that reduces uncertainty.

Even though there is no strategy, one is emerging by default. Road building in urban areas does not seem to be strongly supported except to ease particular bottlenecks. Road investment will take place on orbitals and intercity roads, including the widening of motorways and the construction of new toll roads with some private sector involvement. New rail investment is likely to update existing lines through the introduction of new rolling stock and electrification. New underground lines may be constructed in Central London and there are some 20 proposals for light rail transit in other cities, some of which have already been started (for example in Manchester and Sheffield). Again, the private sector is being encouraged to make some contribution and to take over the operation of services on the publicly provided infrastructure. Ironically, many of these capital schemes are similar to those promoted in the 1960s when there was a similar period of economic growth.

The new realism (Goodwin *et al.*, 1991) identified the publication of the Department of Transport's *National Road Traffic Forecasts (Great Britain)* (Department of Transport, 1989*a*) as a watershed. It was forecast that economic growth and existing trends would result in traffic levels increasing by between 83 per cent and 142 per cent from 1988 to 2025. This doubling of traffic would mean that whatever road construction policy was adopted, congestion would increase and that the key determinant was now demand management. There was no alternative policy as it was now no longer possible or desirable to build roads to match the expected increase in demand. It now became essential to devise policies which would reduce demand to levels commensurate with available supply through a range and combination of policy levers. Five common themes have been identified

- improvements in the quality and scale of public transport;

- traffic calming in residential and central locations, and tilting the balance in favour of pedestrians;

- advanced traffic management systems;

- road pricing;

- assessment of the need for new road construction must follow from a consideration of how much traffic it is desirable to provide for, not how much is required to meet demand.

Two basic questions remain unanswered here. One relates to the desire to reduce the total amount of car use and to achieve an appreciable transfer to other modes (Goodwin *et al.*, 1991). All recent evidence suggests that modal switching is in the reverse direction and that public transport has lost much of its patronage to the car. Such a switch back to public transport may not be feasible in capacity terms, let alone achievable in policy terms. More likely would be the continued use of the car, but to more local destinations and even more complex sequencing of trips. This would allow the same activity participation rates but with less travel distance.

The second, closely related issue is that lifestyles have been adopted which are car dependent. Action is also required on the planning front to ensure location decisions by individuals and companies, together with the availability of facilities, are within closer range of each other. This suggests a much more interventionist planning system that is directive in its approach. Alternatively, the price of land must be raised in locations which are considered to be inaccessible so that development can be 'encouraged' in accessible locations. Accessible in this context means in close proximity to all people so that travel distances can be minimized. The unresolved issue here is that even if travel is more expensive and trip distances are reduced, what will people choose to do with their free time? In the past, increased leisure time and faster travel have resulted in more remote destinations being selected in preference to local destinations. The 10-minute walk is replaced by a 10-minute drive. Underlying the new realism is the essential need to convince the general public of the benefits in financial, time, and social costs, and in environmental terms of using local facilities and only travelling long distance by car or air if it is essential. In the UK, there is little sign that this attitudinal change is taking place, but in Continental Europe it does seem that people are prepared to accept traffic calmed areas and reduced speeds by car.

Related to the realization that forecast demand cannot be met by new road construction is the concern over the forecasts themselves and the *forecasting process*. These forecasts have been essential for the proper allocation of public monies. This covers the ranking of possible new road schemes and helping to decide on the total budget for roads in competition with other areas of public expenditure. Scheme forecasts are mean estimates of traffic volumes on a particular project under consideration. They are obtained from estimates of existing levels of traffic on the road network, and national or local growth factors are then applied to estimate future levels of traffic. A model is used to allocate traffic to individual routes (National Audit Office, 1988). In some cases additional trips expected from new land-use developments are added to the traffic growth, but other forms of traffic generation and redistribution are not normally included.

In the 1970s the criticism was based on the forecasts being too high and there was too much capacity being designed into the system. More recently, the reverse has been true with new roads reaching their design capacity almost as soon as they are opened – 'the M25 effect'. Forecasts of national traffic growth over the last decade have been too low. Part of the explanation is the growth in GDP which took place between 1982 and 1987 (+18 per cent) and the consequent growth in car ownership, and the reductions in the real costs of petrol, and the impact that these two factors have had on car use (figure 1.1). More fundamental though has been the restructuring of industry, the move out of the city, and the changes in lifestyle which have taken place over the last decade. All these changes have been facilitated by the car and patterns of activity have evolved which are now car dependent.

The government claims that 50 per cent of road schemes have been within 20 per cent of actual forecast levels. 'This is a reasonable record given the inherent difficulties in forecasting, and the number of years, typically six or seven, between the time of the forecast and the outturn' (Department of Transport, 1991*a*, para 3.4.6). There is no pattern as the forecast is just as likely to be an underestimate as an overestimate with an average error of 28 per cent in England (19 per cent in Wales) (National Audit Office, 1988). In the North, the North West and the West Midlands traffic flows in most schemes were well below those forecast. In the East and the South West the reverse was true. At the individual scheme level the variation was even more marked. One conclusion could be that the models are being mis-specified with the salient variables being omitted. Another explanation could be the level of generated and distributed traffic associated with new road construction, suggesting that the trip matrix is not fixed. Certainly, an increased attention to sensitivity analysis and the costs of being wrong would help in ensuring that the allocation of public funds to roads is carried out on the best available analysis (House of Commons, 1990). It also suggests that the degree of weight placed on the latest forecasts (Department of Transport, 1989*a*) should be modified in light of the inaccuracy in past forecasts and the fact that no significant changes in methodology have taken place. The new realism among transport experts may also become modified as the forecasts are again proved to be too optimistic or too pessimistic.

The experience of *regulatory reform* in the bus industry and the privatization of many transport enterprises have brought benefits, but it is unclear whether the services provided are more efficient or more competitive. Certainly, productivity has increased, subsidies to the service have been reduced, and most companies are still in operation. But subsidies to the user have increased, fares have increased, and patronage has fallen by a greater rate than trends might suggest.

In a deregulated market economic, efficiency is promoted either through many operators competing within the market or through the maintenance of conditions of contestability, principally through freedom of entry and exit to and from that market. The main objectives of existing operators have been to maintain their monopolistic or oligopolistic control of that market both through anticipatory action and the erection of barriers to entry including collusion with other operators, and through predatory action when entry does occur (Banister, Berechman and De Rus, 1992).

Natural monopolies if they exist may create barriers, threaten retaliation and violate the assumptions of ultra free entry to the market. Only if real competition is possible will entry actually occur and this would lead to the internal structure of the industry evolving towards competition. High barriers lead to high market shares, concentration within the market and higher than normal long-run profits. Conversely, one operator can be given limited monopoly rights to run services on a particular route. Competition takes place off the road in the tendering process and not on the road through direct competition.

Deregulation has helped in reducing barriers to entry and in allowing competition to take place, but there are many strategies that the incumbent operator can pursue to maintain market position. The industry has restructured itself into larger units through mergers and the formation of holding companies; costs have been significantly reduced, and the network has been rationalized with a clearer indication of cost allocation to particular routes. Consolidation within the industry has taken place, and innovation through improved vehicle quality and minibus operations have all constituted the main benefits of deregulation in the bus industry. The losses have been the fares re-adjustments, and a period of intensive instability with poor information and confusion for the customer. Subsidy has been reduced but patronage has also declined significantly on urban bus services. It is still unclear whether bus operations are more efficient in the private sector, whether the market is contestable, or whether the bus industry is a natural monopoly.

For other parts of the transport industry, the benefits of regulatory reform and privatization are clearer. Associated British Ports, the British Airports Authority, British Airways, some of British Rail's subsidiaries, the National Freight Corporation, and Jaguar have all benefited from their return to the private sector. Efficiency and productivity have improved, with profitability and return on investment being clear measures of performance. Greater autonomy and control has been given to management, with clear corporate objectives and accountability. These arguments may seem attractive, but Kay and Thompson (1986) have argued that the case is less than clear. The policy of privatization does not have a sophisticated rationale, rather it lacks any clear analysis of purposes or effects, hence any objective which seems achievable is seized as justification. The outcome is that no objectives are effectively attained, and in particular that economic efficiency has systematically been subordinated to other goals, in particular political goals.

This argument is well illustrated by the current debate over the privatization of the railways where there is considerable political impetus for action, but where the alternative forms of privatization are not attractive. It is very difficult to maintain a competitive market for rail operations on a complete network basis. Parts of the network can operate within the private sector (for example freight), and another option could be to separate track ownership from the operation of services.

Legislation has been introduced in 1993 to introduce this last favoured option. A separate track authority (Railtrack) will charge private and public operators to run services, and a new franchising authority will determine the contract requirements for each route. The first six major lines for franchising have been identified and if there is private sector interest, British Rail monopoly will be broken early in 1994. Apart from the exact form of privatization, two other problems are endemic. The railways have always lost money, despite extensive network and service rationalization, and the use of property sales to finance investment. It seems likely that this position will not improve. The second problem seems to be that if any transport service is a natural monopoly, it would be the railways (Banister, 1990). The political arguments have dominated both the economic and planning arguments in the 1980s.

The *environmental* concerns of the 1990s may be the key to change with the broadening of concern from local issues (for example noise, severance and visual intrusion) to the global impacts (for example acid rain, pollution and greenhouse effects). Global warming is now taking place and concepts such as sustainable development and the principle that the polluter pays are high on the political agenda. The growth in the demand for transport is proceeding at unprecedented rates, both in the UK and the world as a whole.

Difficult choices have to be made. To accept the arguments for sustainability will mean a reduction in consumption and a much higher price for travel (Banister, 1993). It will also mean striking an appropriate balance between economic growth, the distribution of that growth and the environment. Environmental issues only form part of the political spectrum, and politicians have to balance economic growth with sustainable growth. Transport revenues form a significant part of total Exchequer revenues, and by raising the cost of transport to the user to reflect the full social and environmental costs of travel may affect Exchequer revenues. Even if the price elasticity is low and people are prepared to pay more for travel, expenditure patterns in other sectors may be affected causing in turn a fall in demand, reduced growth and possible unemployment.

In 1990 the Department of the Environment published a White Paper, *This Common Inheritance* (Department of the Environment, 1990) which claimed to present the first comprehensive review of every aspect of Britain's environmental policy (table 4.6). However, the options for transport seemed to be limited in their scope as they primarily related to reductions in energy consumption. The environment must be interpreted more broadly, and the environmental effects of transport influence all our lives in a variety of ways. Here are just two examples of the complexity of the interactions and the breadth of the transport and environment links.

In the past demand forecasts have been made for traffic and networks have been defined to meet that demand. It has now been realized that it may not be socially efficient, or desirable or possible to meet unrestricted demand and so restraint and management have become key concerns of transport planners. Alternative means of forecasting are being used such as scenario building, where alternative strategies are tested against possible future scenarios (Department of Energy, 1990). Alternatively, simulation studies have been developed to establish energy efficient forms of urban settlement patterns (Rickaby, 1987). In most cases these studies have taken one or two sectors (transport and land use) and examined environment in terms of one variable (energy). Yet no one has established the conditions under which a settlement

TABLE 4.6. Transport policy options proposed in the White Paper *This Common Inheritance* (Cmnd 1200).

### 1. VEHICLE STANDARDS

Tax incentives to reduce emissions
Standards for cars
Standards for heavy diesel vehicles
Road planning
Public transport

### 2. TAXATION

Vehicle excise duties
Tax on company cars
Duties on petrol and diesel

### 3. CAR MANUFACTURERS AND CONSUMERS

Pressure on EC to tighten fuel efficiency standards on large cars
Support for EC proposals to require catalytic convertors on all new cars by the end of 1992 (deferred for one year)
Government encouragement for fleet operators and individuals to buy cars fitted with catalytic convertors
Extension of MOT testing to cover emission of carbon monoxide, other polluting gases and possibly noise
Stricter enforcement of speed limits and possible lower limits in the long term

### 4. HEAVY DIESEL VEHICLES

Pressure for stricter EC standards on particulate emissions from new vehicles
Support for US style restrictions on $NO_x$ emissions by 1996/97
Possible application of tighter standards to existing vehicles which may necessitate retrofitting
Extension of MOT tests to include emissions of polluting gases, smoke and possibly noise, with proposals to increase the frequency of roadside tests
Encouragement to local authorities to control lorry movements in urban areas
Consideration of environmental benefits in the award of grants for rail or water-borne freight schemes

### 5. ROAD PLANNING

Reduction in congestion through the creation of 'red routes'
Construction of by-passes and trunk roads, in conjunction with local authorities where appropriate
Support for local authority parking schemes
Research into design to reduce road noise

### 6. PUBLIC TRANSPORT

Continued financial support for British Rail and London Transport
Local authority action on priority bus lanes and traffic lights
Increased provision of cycle networks
Research into noise from public transport

*Source*: Based on Coopers and Lybrand, Deloitte (1990).

pattern is energy efficient, let alone environmentally efficient. Transport factors have to be balanced against other energy costs such as the energy used in the construction and maintenance of the infrastructure (including buildings) together with the costs of space heating and ventilation. The energy factors have to be balanced against the economic costs of development, the availability and price of land, and labour costs. The qualitative factors which make up the environment add to this complexity. The research issue here is whether it is worthwhile to unravel these complexities and measure the interactions, or whether it is more sensible to take a modest perspective and examine policy questions individually.

The second example is that of health and stress as it relates to transport. In recent surveys carried out at the Marylebone Centre Trust (Tennyson, 1991) it seems that transport is now a major cause of stress both for users of the system and for those who live near to major transport routes. On the one hand, transport raises people's anger and aggression through such factors as the delays in traffic and actions of other drivers, whilst on the other hand transport creates fear and worry through the difficulties of crossing roads or travelling in very crowded conditions. Little research has been carried out on the effects of this on people's health, their absenteeism from work, their performance at work and their job satisfaction. In addition to the stress factors, public transport and certain spaces (tunnels and some roads) are perceived to be unsafe, and questions of security can help explain why some people are reluctant to travel after dark on their own (Atkins, 1990).

There is a wide range of 'solutions' (table 4.7), yet little progress has been made to reduce the emissions from transport (apart from lead in petrol) or to reduce the consumption of fossil fuels. It is expected that emission levels and consumption of oil will continue to rise, with any benefits from increased efficiency being outweighed by the continual growth in traffic. If the UK is to achieve the target set by the Climate Change Convention (Department of the Environment, 1992) to return the emissions of each greenhouse gas to 1990 levels by 2000, action is required in all sectors. Yet in transport, the dilemma is clear as there seem to be no obvious means to reduce emissions and energy consumption levels which are politically acceptable. The possible environmental benefits have to be balanced against the need to reduce unemployment, to maintain individual choice, to increase the competitiveness of industry, and to maintain or increase Exchequer revenues.

These five issues present government with major choices, but all the problems would be much easier to resolve if there was a clear statement on transport policy which would then be pursued consistently over a period of time. Fifteen years of Conservative government have witnessed radical change in government policy, and transport has been the sector which has been the testbed for many of these innovations. Many of the accepted roles for the state have been questioned and replaced. As with all radical changes, a reassessment is also

TABLE 4.7. 'Solutions' to transport and pollution.

| 1. | TECHNICAL FIXES 1: POLLUTION REDUCTION AND ENERGY EFFICIENCY |
|---|---|

| Pollution reduction technology | Oxidation catalysts<br>Three-way catalysts<br>Catalytic trap oxidizers |
|---|---|
| Improving energy efficiencies | Engine changes (e.g. lean burn)<br>Weight reduction<br>Aerodynamics<br>Other technological modifications (e.g. transmission changes, rolling resistance) |

| 2. | TECHNICAL FIXES 2: ALTERNATIVE FUELS AND POWER SOURCES |
|---|---|

Diesel
Electricity
Hydrogen
Alternative power sources (e.g. nuclear power, gas from power stations, renewable sources)
Gas (e.g. Liquefied natural gas, liquefied petroleum gas)
Methanol and ethanol

| 3. | THE ROLE OF THE DRIVER |
|---|---|

Lower average engine size: the vehicle purchase decision
The vehicle replacement decision
Increasing car occupancies
Better driving practices
Better maintenance

| 4. | TRANSPORT PLANNING POLICIES |
|---|---|

| Intermodal shift | |
|---|---|
| Road traffic management | Improving traffic flow<br>Reducing excessive speeds<br>Discouraging car traffic |
| Land-use planning | |
| Other policies | Public information campaigns<br>Encouraging telecommuting |

| 5. | TRANSPORT PRICING POLICIES |
|---|---|

Road pricing
Fuel pricing and taxation policies
Company car tax policies
Vehicle pricing and taxation

*Source*: Based on Transnet (1990).

needed as to whether these ideologically driven policies have been successful and whether they have achieved their objectives. However, the difficulty is that as they have never been part of a coherent and consistent policy, it is almost impossible to determine the criteria on which success or failure can be gauged: perhaps that is the intention!

# Chapter 5

# Overseas Experience

The problems outlined in the previous three chapters are not restricted to Britain. Urban transport planning throughout the developed world is now faced with a series of paradoxes and dilemmas which have been neatly summarized by Hall (1985).

1. From the mid-1960s onwards, urban transport planners almost everywhere forsook the ideal of individual motorized mobility for all. They substituted a combination of urban traffic restraint and the promotion of good public urban transport.

2. Their change of mind (or heart) was fortified in the 1970s, first by the limits of growth arguments contained in the Club of Rome report of 1972, second by the living manifestation of that argument in the great energy crisis of 1973–74, and third by the appreciation of ecological impacts throughout that decade and into the 1980s.

3. The first two of these influences are now much diminished in force. Soon after 1973–74, energy prices resumed the broad secular downward courses (in relation to general price levels) that they have manifested, with only brief perturbations, throughout the entire post-World War Two period. The ecological argument in contrast looms stronger than ever, though its precise significance for transport policy is not always clear and is certainly debatable.

4. Meanwhile, throughout the 1970s, there was a clear and very general tendency towards dispersal of population – and, in some cases, economic activity – from cores to peripheries in urban areas. There are wide variations in the extent and the time of this process as between one country and another, but it may be that these merely represent points along a common trend line. If the pattern continues, it can only strengthen the use of the car and weaken the position of public transport authorities.

5.   In any event, the financial position of these public transport authorities has in general deteriorated because of the growth of car ownership and use. The resulting central and local governmental subsidies are now being challenged because of the pressure of the recession on public expenditure.

6.   There has recently been increasing interest among transport planners in a disaggregated approach that focuses on the needs of individual household members within a time space framework. This has thrown sharp light on a problem that rising household car ownership may only conceal, namely a lack of car access on the part of members – even a majority of members – of that household. This may be exacerbated by the deteriorating quality, or increasing cost, of the public transport alternative.

Hall (1985) argues that these crucial paradoxes are not addressed either by transport professionals or by politicians. Even at the supra-national level, the policy dilemma is apparent in the European Community's desire for a Common Transport Policy. The principles contained in the Treaty signed in Rome (25th March 1957) and the Single European Act signed at Luxembourg (12th February 1986) and at the Hague (28th February 1986) declare the necessity of free competition and of the free circulation of persons, goods, information and ideas for achieving cohesion within the European Community nations.

Over the last five years the EC has passed two important Regulations (Reg 4056/86 and Reg 4048/88) which establish the objectives for a Community policy of transport infrastructures in the medium term together with their means of financing. The objectives include the

*   elimination of bottlenecks;

*   integration of peripheral areas;

*   reductions in costs of transit through third countries;

*   upgrading of connections in land-sea corridors;

*   establishment of high level service links between the main cities including high speed rail.

The criteria for investment cover Community wide benefits, socio-economic profit and increased coherence within the Community. It seems that the Community interest is defined on the basis of the levels of intercommunity traffic, not the creation of a homogeneous and balanced network. Growth is more important than distributional questions associated with improving the accessibility of peripheral regions. This position has been modified by more recent statements made by Jacques Delors (17th January 1990) to the Commission, where the dual focus of the development of Europe-wide networks and a common transport system has been promoted. The free movement of services is balanced by a concern over comparable conditions of competition, optimum working conditions, and the environmental imperative. There needs to be a balance between liberalization and harmonization. In the past, local and regional level decisions have been too introverted and related to specific situations, but there now seems to be a greater vision for Europe and a stronger role for planning.

The enormous difficulties in determining a Common Transport Policy for an ever expanding EC and for a greater Europe should not be underestimated. There are fundamental differences between individual member states' cultural traditions, the means of intervention and regulation, attitudes towards competition, and the whole issue of investment and financing in transport. These issues are returned to in the later debate on the key issues facing transport planning in Europe (Chapter 7), where the role of the EC in infrastructure investment is discussed. Here, the argument revolves around the different traditions of transport planning which have evolved in particular national contexts.

In Britain, there has been a move towards greater efficiency through the use of the market mechanism and a reduction in levels of public support for transport. In other countries (for example France, Germany and the Netherlands), transport is seen as part of a wider social policy where allocative efficiency is more important than cost efficiency. These different traditions are highlighted here with the European traditions being compared to those developed in Britain and the USA. The traditions of transport planning and urban planning in Britain and the USA are directly comparable, even though the policy context both politically and economically may have been different (Chapter 2). However, in the 1980s even these converged with two strong conservative governments.

## Transport Planning in the USA

Conventional thinking in the 1950s and 1960s was that roads should be built to accommodate the era of universal motorization. Car ownership in all advanced industrial countries was rapidly increasing and saturation levels would be reached where some 90 per cent of all adults (17–74 years of age) would have their own car. This would give a car ownership level of 650 cars per 1,000 population. The challenge for highway planners was to develop a network of all-weather highways to meet that expected demand. In the USA, a series of standard procedures was developed for application to all urban areas – *Better Transportation for Your City* (1958).

The methods used were aggregate in scale

- Fratar method for trip generation;
- gravity models for trip distribution;

- traffic diversion curves, based on empirically derived travel time ratios;

- shortest path assignment algorithms.

They were designed to place urban transport planning on a more systematic basis. The seminal studies carried out in the Detroit Metropolitan Area Traffic Study (1953–55), with many of the calculations done by hand, and the Chicago Area Transportation Study (1955) were a tribute to the pioneering spirit of the time, and the enterprise of Douglas Carroll who directed both these studies. The optimistic view of that time is captured in the two benchmark books by Meyer, Kain and Wohl (1965) and Creighton (1970) on urban transportation planning.

In March 1962, a joint report to the President on urban mass transportation stated:

> transportation is one of the key factors in shaping our cities. As our communities increasingly undertake deliberate measures to guide their development and renewal, we must be sure that transportation planning and construction are integral parts of general development planning and programming. (Weiner, 1985)

The importance of this statement is that it marked a switch of resources from highway construction to urban mass transportation, and there was a requirement (in the Federal Aid Highway Act, 1962) to develop programmes which properly coordinated and evaluated the impact of future development of the urban area. The Bureau of Public Roads set up the basic elements of the '3C Planning Process' (continuing, comprehensive, cooperative) for which inventories and analyses were required to cover:

1. economic factors affecting development;
2. population;
3. land use;
4. transportation facilities including those for mass transportation;
5. travel patterns;
6. terminal and transfer facilities;
7. traffic control features;
8. zoning ordinances, subdivision regulations, building codes etc;
9. financial resources;
10. social and community value factors, such as preservation of open space, parks and recreational facilities; preservation of historic sites and buildings; environmental amenities; and aesthetics.

The 3C Planning Process led to the development of standardized procedures, computer software and training courses for transport professionals, and by 1965 all 224 urban areas had started their urban transport planning process (Weiner, 1985). For the first time federal grants were available for mass transit construction costs, and for the acquisition of facilities and equipment (Urban

Mass Transportation Act, 1964) provided that those costs could not be paid out of revenues.

It was also during this period in the turbulent 1960s that two other issues came to prominence. Firstly, there was a reaction against the highway construction programmes, initially in San Francisco, but soon mirrored across the Western world. The premise on which highway planning was predicated, namely that the demand for car travel had to be met by increased road capacity, was shown to be false. The impact of the Buchanan Report in Britain (Chapters 2 and 3) has already been noted. The equivalent report in Germany (Hollatz and Tamms, 1965) argued that the growth in car ownership and use had to be met through the construction of orbital and ring roads, combined with restraint on car use in city centres and heavy investment in public transport (Chapter 5). In the USA, Meyer, Kain and Wohl (1965) tested a series of hypotheses through a range of exhaustive empirical investigations. These hypotheses were:

- the decline in city populations and densities may be attributed to poor public transport;

- a rail-based solution would be cheaper than a road based one;

- travellers would be willing to switch to public transport if the quality was improved;

- urban transport shapes the city.

In the 20 years after the war, public transport use in the United States declined by 64 per cent (1945–1963), but route miles of railway increased by 2 per cent and route miles by bus increased by 30 per cent. The situation in Britain is slightly different. Here, peak public transport use was in 1952, not in 1945. The decline over the subsequent 30 years has again been over 60 per cent, but route miles have also been cut (by 20 per cent). The conclusion reached is that if high levels of mobility are desirable and car ownership continues to rise, then the choice seems to lie with the construction of extensive networks of new roads in urban areas unless travellers can be diverted back to public transport.

The second issue, again picked up in the Meyer, Kain and Wohl book (1965), was the lack of coordination between different government agencies, principally relating to housing and transport. The Department of Transportation was created (1966) to match the new Department of Housing and Urban Development (1965). The end of the 1960s marked a period of uncertainty in the use of the systems approach to transport analysis as models had not matched expectations, particularly at the local level. It was also realized that the large-scale aggregate approach, together with the well established procedures for data collection and analysis, did not cover the broader social goals or behavioural factors.

Different approaches seem to have been adopted by Buchanan and Meyer, Kain and Wohl. *Traffic in Towns* attempts to examine a range of options through a series of practical studies that represent the spectrum of urban areas in Britain. The American study of *The Urban Transportation Problem* is not so comprehensive and focuses on a set of particular issues in a systematic manner, such as the relationship between housing and transport, and race and transport. Perhaps the most important contrast is that Buchanan's options examine different levels of redevelopment and the notion of environmental areas around which traffic would be diverted, whilst the focus in Meyer, Kain and Wohl is on the modal split question and the balance between public and private transport. Both issues are encompassed in Appleyard's concept of 'liveability' and environmental capacity (Appleyard, 1981) where it is suggested that there is an inverse correlation between traffic volumes and neighbouring (liveability). The essence of the argument is to make the car more amenable to the city and *not* the city to the car. Appleyard (1981, pp. 130–31) claims that the concern in *Traffic in Towns* for the environmental impacts of traffic predate official American interest in the problem by about five years.

On both sides of the Atlantic, the early 1970s marked a watershed in thinking as high oil prices, global recession and increased unemployment caused political and economic uncertainty. Priorities switched from road construction to the means to improve capacity and safety through traffic operations (TOPICS), and the requirement to hold public hearings on proposed highway projects. The continuing urban transport planning process should cover five elements (Instructional Memorandum 50-4-68)

- surveillance – monitoring change;

- reappraisal – review and update proposals;

- service – assistance in implementation;

- procedural development – improved analytical skills;

- annual report – wider communication with the public and officials.

Even though thinking on the environment may have come later to the USA, action has been far more decisive than in Britain. The pioneering National Environmental Policy Act (1969) required Environmental Impact Statements for major federal actions and for all legislation, and this in turn necessitated a systematic interdisciplinary approach to planning and decision-making. The Council on Environmental Quality would implement the policy. It was only in June 1985 that the European Community issued the Directive on the Assessment of the Effects of Certain Public and Private Projects on the Environment, including the construction of motorways and express roads – interpreted in Britain as trunk roads over 10 km in length (Article 4(1)), and other roads or urban development projects at the discretion of individual member states (Article 4(2)). The Directive took effect from July 1988 (Chapter 4).

The 1970s became the decade of public transport, with significant increases in capital and revenue support. The Urban Mass Transportation Assistance Act (1970) gave priority to schemes for helping elderly and handicapped travellers whilst all highway projects had to give a full analysis of economic, social and environmental effects (Federal Aid Highway Act, 1970). The main innovation in analysis methods was the use of behavioural and disaggregate models, together with a realization that short-term, quick-response, simple methods were most informative for decision-makers. The switching of resources continued with the Federal Aid Highway Act (1973) which permitted the use of highway funds for urban public transport for the first time. A year later, the national Mass Transportation Assistance Act allowed the use of federal funds for public transport operating subsidy.

San Francisco's Bay Area Rapid Transit (BART) epitomizes many of the debates concerning public transport. Operations began in October 1974 over the 71-mile system, and by 1976 BART was carrying only half the forecast passengers and the operating loss had reached $40 million. These losses compounded the high capital cost overrun of 150 per cent and the high operating costs, some 475 per cent of forecast. Average costs per ride were twice those of the bus and 50 per cent greater than a standard American car (Webber, 1976). One of the main arguments for the investment was the potential for BART to attract travellers from the car, but with only 5 per cent of the peak trips, it has made little difference. The importance of access time to and from the system was underestimated. Webber (1976) concludes 'it is the door-to-door, no-wait, no-transfer features of the automobile that, by eliminating access time, make private cars so attractive to commuters – not its top speed. BART offers just the opposite set of features to the commuting motorist, sacrificing just the ones he values most.'

BART has not noticeably influenced land-use patterns, except in particular locations such as at Walnut Creek. The intention was to create highly accessible locations which would prove attractive to new businesses, offices and housing developments. The localization effects would generate further concentration at these sites as multiplier effects took place. The mistake was the expectation that the railway would make any impact on accessibility in an urban area which already has a highly developed road network. The technical planning process set no formal goals, looked at no alternatives, and made no formal evaluation until late in the process. The problem was to alleviate traffic congestion coupled with the protests over freeway construction, and the mechanisms for public consultation were weak (Hall, 1980).

The third factor was the availability of federal capital to match locally raised capital. With the setting up of the Urban Mass Transportation Administration within the Department of Transportation (1968) and the subsequent

commitment of federal aid to public transport investments, huge new investment projects could now be undertaken. The funding was concentrated in 1600
miles of new urban rail lines including Washington, Los Angeles, Baltimore,
Detroit and Atlanta. It seemed that during the 1970s there was a genuine
feeling supported by local politicians that rapid transit systems could get
people to switch from car to rail. New technology and heavy investment in
urban rail would provide the solution to urban congestion.

In retrospect, the arguments may seem naive, but given the range of possible
alternatives there may have been little option. The path followed in Britain
was somewhat different. Heavy capital investment in roads was curtailed in
the 1970s, but there was little diversion of resources to capital investment in
urban rail systems – with the exception of the Tyne and Wear Metro in
Newcastle (1974–1984), the extension of the Victoria line in London (1971),
the extension of the Piccadilly line to Heathrow (1977), and the new Jubilee
line in London (1979). In both countries, the short-term response was a heavy
commitment to further subsidy for both bus and rail systems.

Transport planning in the USA in the 1980s moved towards the managing,
maintenance and replacement of existing facilities, together with an extensive
analysis of the means to limit the use of the car in the city. Apart from the
problems of physical capacity, there was an increasing concern over energy
consumption in transport and the growth in transport-related levels of pollution. All these factors helped focus attention on the car. Urban policies were
also directed at the means by which city centre decline could be reversed. The
confidence of long-term planning, characteristic of the previous two decades,
was replaced by a more cautious pragmatic incremental approach to planning.

The key to this new approach has been the decentralization of decision-
making and a reduction in federal intrusion into local decisions (Weiner,
1985). The complexities of planning as a social process with important environmental and ecological dimensions, together with tighter fiscal constraints
and increased private sector involvement, all make it more appropriate for
devolved decisions to be taken. The rigid large-scale physical approach to
planning was not suitable for smaller-scale decisions which required local
flexibility. Coupled with this decentralization was a necessity to reinvest in
upgrading the existing road infrastructure. The Surface Transportation Assistance Act (1982) increased the '4R Programme' – resurfacing, restoration,
rehabilitation, reconstruction – by raising user charges from 4 cents to 9 cents
a gallon of petrol. The new Urban Transport Planning Regulations (1982)
retained the requirement for a Transport Improvement Program (TIP) including a Unified Planning Work Program (UPWP) for all areas with a population
over 200,000, but the process was to be self-certified by the state and the
Metropolitan Planning Organizations (MPOs).

However, at the local level there is a separation of transport and planning
functions, and a shortage of available expertise. The land development process

is mainly a private sector operation and there is little expectation that a plan will be implemented. Even zoning is carried out to meet expectations rather than reality, and much planning is merely modifying existing plans or carrying out rezonings (Deakin, 1990*a*).

Underlying all transport planning problems of the 1980s and early 1990s is the key problem of congestion and the prospect of total gridlock in city centres and suburbs. Demand patterns are now so complex and suburbs are no longer dormitory locations for commuters working in the city centres. Suburbs in the main metropolitan areas are now the places where shopping centres are located and the recent growth has been in business parks and the 'horizontal' office blocks (groundscrapers). They are attractors and generators of traffic, yet the evidence on whether trip distances have increased as a result of suburbanization is less than clear (Cervero and Hall, 1990; Gordon and Richardson, 1989). The Californian Transportation Commission (1987) has estimated that congestion results in 75 million hours lost annually within the State, and about half of this delay is caused by non-recurring incidents such as accidents, lane closures and unpredictable events.

Transportation Systems Management (TSM) has been extensively used to shift demand over space and time through more flexible working practices and through the use of vanpools and carpools. With the cutbacks in the federal budgets, the private sector has become more involved in funding roads, with their contribution now reaching about 20 per cent (Cervero and Hall, 1990). Projects range from local improvements to a new generation of toll roads financed by bonds and toll revenues. Traffic volumes of 50,000 vehicles a day are needed to justify toll financing so that interest charges and maintenance costs can be paid (Deakin, 1990*b*). Federal aid is now also available at a 35 per cent matching level for a small number of toll road demonstration projects, but it is often these supplementary funds and other guarantees from developers and government (for example development impact fees and tax benefits) which make the decision worthwhile.

Other forms of road pricing seem less attractive in the USA and there is little evidence of peak pricing except on the Washington Metro and on certain bridges (for example San Francisco Bay Bridge). But even here, there is only an implicit acceptance of pricing as high occupancy vehicles are exempt and as there is no difference in rates between the peak and off-peak. Parking is usually provided at a low cost except in the central core and zoning standards often overestimate the amount of parking that developers should provide. It is likely that there will be a huge resistance to road pricing in the USA. Comprehensive regional planning seems as far away as when the Regional Planning Association of America called for it in the 1920s (Cervero and Hall, 1990).

Superimposed on the threat of gridlock is that of the environment. It is here that some of the most interesting innovations have taken place as land use is being controlled to reduce the rate of growth in traffic. In Southern California, Proposition U reduces the allowable development on most land zoned for commercial development. The measure halved allowable development along certain main roads from 3:1 to 1.5:1 (the floor area ratio). The Traffic Reduction and Improvement Program (TRIP) allows the Council in Los Angeles City to designate any neighbourhood a 'traffic impact area'. For each area of transport, a specific plan is developed including the requirement for the developer to calculate the trips generated by the proposed development (from the *Manual of Trip Generation Rates* produced by the Institution of Transportation Engineers). The developer must then mitigate the effects of those trips by reducing the afternoon peak hour trip generation of the project by at least 15 per cent. Failure to achieve this reduction results in a fee being paid into a trust fund and the monies can be used to pay for improvements in the impact area's transport plan (Wachs, 1990). The introduction of parking charges at workplaces has been the most effective way to meet these targets with wages being raised to compensate for the loss.

Regulation XV imposed by the Southern California Air Quality Management District (SCAQMD) requires that each employer with 100 or more employees has to ensure that their workforce achieves an Average Vehicle Ridership (AVR) threshold for journeys to work. The AVR is the vehicle occupancy rate and relates to all employees and all cars. So the impact of the use of public transport, bicycle and walk modes will all be positive. The actual levels of AVR depends on the city location with the Los Angeles CBD value being 1.75 and the suburban value being 1.3. Again, fines are imposed on non-achievement of the targets, and it was estimated (Atkins, 1992) that in August 1991 some 209 employers out of a total of 6,900 were in violation. In each of these cases, the problem has been addressed through the workplace, either in terms of reducing the impact of new developments or in reducing the numbers of cars used for the journey to work. However, the environmental and congestion problems created by the car do not only relate to the work journey, but to a multitude of other activities which are much harder to control (for example social and recreational trips). Work journeys in both the USA and Britain are relatively stable in number, and form a diminishing proportion of the expanding travel market.

COMMENT

The basic philosophy behind land-use and transport planning in the USA is very similar to that in Britain. In the early stages with the development of the systems approach, Britain followed the theoretical and analytical path pioneered in the USA, but more recently with the increased use of disaggregate and behavioural methods that interaction has been more balanced. In both countries there is at present a pause in the development of methods and their application in planning practice. This pause is in part due to the move away from large-scale strategic planning, but more particularly to the growing interest in smaller-scale localized applications of particular analysis methods to immediate problems. The concern over comprehensiveness and order is being replaced by one of selectivity and specialization.

The main difference between approaches has been the trend in the USA towards decentralization, whilst in Britain increased centralization has taken place. In the USA, the states have always had more power over local decision-making and the raising of local taxes than have the counties in Britain. The 1980s have been characterized by reductions in public budgets at both the national and the local levels, yet in the USA there is more opportunity to raise cash locally. Tax hypothecation is permitted in the USA through the levy on petrol prices, and the private sector seems to have been more actively involved in the new generation of toll road construction. The planning system in Britain is still more restrictive than that in the USA, and as such the opportunity for private investment in roads has been limited. The possibility of federal support for capital expenditure in rapid transit systems and toll roads, together with advantageous development benefits, have all resulted in greater private sector activity in the USA (Deakin, 1990b). Comparable schemes in Britain are more limited and are much more closely linked in with the business cycle in the building and construction industries.

The other main difference between the two countries has been the continued subsidization of public transport in the USA, whilst in Britain the government has managed to reduce subsidy levels, principally through regulatory reform in the bus industry and through the imposition of severe productivity targets for the railways. Transport professionals have never really been at ease with the increased politicization of transport issues. As Wachs (1985a) comments, subsidy is commonly seen as inefficient and analysis has been 'so uniformly critical of American public transit policy for so long, yet the trend towards greater subsidies, more inefficient fares structures, and continuing shifts in service towards the lower density, higher cost markets continues.' All these studies have been ignored as the technical arguments do not address the political criteria which are dominant in the process of resource allocation. Equity, in political terms, is the share that each constituency gets of the budget and not the extent to which those who benefit from a service actually pay for it or actually need it. Efficiency is the quantity of service produced by each dollar of subsidy, not the fare box revenues or measures of productivity. The gasoline tax seems to summarize the difference between political and technical decision-making. The gasoline tax is a user tax, yet 20 per cent is allocated to transit funding, but the majority of Americans never use buses. So the tax is in effect being paid by motorists for the non-use of buses.

## Transport Planning
## in the Federal Republic of Germany

The main priorities in Germany have been the reconstruction of the infrastructure and cities destroyed in World War 2 and the revitalization of economic production. The early 1960s brought an increase in car ownership, but it was only in 1965 that Germany had about the same number of cars as Britain (8.5 million). The tradition in transport planning was to collect all the relevant data through traffic counts, origin and destination surveys and traffic densities so that forecasts could be made. The techniques and practice developed in the USA were used as German traffic engineers had limited experience themselves (Hass-Klau, 1990). Even though public transport was still the main mode of transport in large cities there were arguments about whether the trams should be replaced as they interfered with the free movement of the car. The free market ethos extended to road transport and the journal *Verkehr und Technik* wrote in 1953 'constructing roads is not only an issue for the car users; it is an issue for the whole nation.'

The standard book used at this time was Leibbrand's *Verkehr und Städtebau* (1964). It favoured the separation of traffic as this allowed faster speeds and greater safety, and argued that the car was the salvation of the city. Free access to the city centre would allow growth in the urban economy and the maintenance of prosperity. It seems that the Germans were more car oriented in their thinking and this was reflected in their road building programme. However, even at that time concerns were arising over the implications of the continued programme of road building. The second major report *Die Kommunalen Verkehrsprobleme in der Bundesrepublick Deutschland* (Hollatz and Tamms, 1965) brought together the leading German researchers to review the problems of traffic and their conclusions were still clearly in favour of the car

> it cannot be considered to restrain private motor transport: a rational design for urban traffic was needed. A sensible division of different transport modes was necessary, which implies traffic restraint in particular locations.

The report recognized that urban structure had been influenced by the car and proposed a system of tangential and orbital routes around cities. Traffic had to be controlled in city centres, principally through parking restrictions, and public transport should have a separate right of way.

Strangely, there seems to have been little interaction between the Germans and the British as the Buchanan Report *Traffic in Towns* was being prepared at the same time. Similarly, in the USA there was a major shift in thinking with the establishment of the 3C Planning Process (1963) and the Urban Mass Transportation Act (1964) which allowed the use of federal funds for capital projects in public transport for the first time. In 1969, a federal law in Germany established a 15-year plan for the completion of the federal highway system.

Prior to 1967 public transport did not receive any funds from the federal government, and by 1965 it had reached a low point in terms of efficiency and the numbers of passengers carried (Hall and Hass-Klau, 1985). Even at this early stage, a change in philosophy was detectable as urban planners were active in the pedestrianization of city centres – over 60 cities had introduced schemes (Monheim, 1980), including Bremen with its system of four traffic cells where movement by car could only take place within each cell and not between cells. The Federal Act (1967) increased petrol tax (*Mineralölsteuer*) by 3 pfennig a litre with 40 per cent of the revenues being used to finance public transport investment. This proportion was later increased to 50 per cent, but by 1989 it was back to 40 per cent. In Germany, the road lobby has always been particularly powerful and apart from restricting the amount of money transferred to public transport from the petrol tax, they also managed to get a dense network of interurban motorways approved so that 85 per cent of the population would be within 10 km of the system.

The end of the 1960s was also marked by student revolts which (as in France) changed politicians' attitudes towards urban renewal and public transport. It was also the first time that the Social Democrats had come to power (1969), with the expectation that liveable cities and a more socially aware society would be created. The dominance of the car in that society was beginning to be questioned by Dollinger (1972) and Dahl (1972). Again, as with other countries it seems that there was a window of optimism prior to the global events of 1973, which marked an end, at least temporarily, to the growth in incomes and prosperity.

As in Britain and the USA, Germany made use of the aggregate models of land-use and transport, establishing relationships between people, motorization and mobility, predicting changes based on population trends and developing a five-stage sequential demand model. The desirability of a trip or attractivity of the place was established from the population and social status of the location, and this was followed by the conventional trip generation, trip distribution, modal split and assignment stages (Mäcke, 1964). The transport analysis was then followed by a functional analysis of the transport system and the city as a whole. Regression techniques and gravity models seemed to be the most commonly used techniques, with cost benefit analysis being used to evaluate the alternatives.

Reaction against the use of aggregate models was triggered by Kutter (1972) who questioned the assumption that traffic forecasts could be based solely on objective data from the populations of areas. He argued that travel behaviour must be based on surveys of behaviour and attitudes. The assumption that people wish to travel from places of small 'mass' to places of large 'mass' is incorrect (that is, the notion of gravity does not apply to people's movements) as it postulates the independence of spatial characteristics and human motivations (Paradeise, 1980). With the use of disaggregate models the individual could be re-established as the prime unit of analysis which would allow analysis of constraints, transport being viewed as a derived demand. Such an approach would allow greater transparency and increase the predictive value of models by incorporating structural changes in population. Three types of disaggregate demand models have been developed:

- Econometric models which do not appear to have been developed in Germany. These have mainly been used in the Netherlands and the USA.

- Models based on the identification of homogeneous behaviour of groups of people. These have been extensively used in Germany.

- Situational approach which extends the modelling framework to include decision and learning theory. Again, Germany has been instrumental in the development of these stochastic methods which attempt an understanding of motivations, and the identification and measurement of preferences.

The professional involvement of the traffic engineer has always been a powerful influence on the development and use of techniques, and that professionalism was dominant in the 1960s and 1970s. Even when new methods were being developed, it was still the engineer who was using them. Even when the social scientist became more involved in transport research, there were still institutional barriers to the acceptance of new methods as decision-makers came from the same traditional engineering background.

In 1973, the *Bundesverkehrswegeplan* (the Federal Transport Plan) provided a framework for all transport, including public transport. The catalyst for action had come from Munich where the first rapid rail system had opened (1972), and the extensive pedestrianization of the city centre together with new housing and development on the periphery received world wide attention. Such schemes had been facilitated by the *Städtebauförderungsgesetz* (Urban Development Act) (1971) which promoted the renewal of inner-city areas with extensive federal support and gave active encouragement to stronger citizen participation in these decisions. By 1975, expenditure on public transport was about five times that in 1967, but the Federal Planning Reports at that time were still presenting the same argument as those used in the 1960s. The major reason for inner-city decline and the movement of people out of the city centre was the damaging effect of the car, which will according to the forecasts increase in the future. The conflict between transport and urban structure will sharpen (Hall and Hass-Klau, 1985). Two measures to counter this decline have been the promotion of public transport in city centres and the concentration of through traffic onto major roads to allow traffic free areas in city centres and residential areas. Yet the federal ministry did not allocate investment to support the public transport measures.

As noted earlier, pressure groups in Germany have always had a powerful influence on policy. In the 1960s and 1970s the road lobby maintained the heavy motorway investment programme, but it was the rise of the Greens and the environmental movements in the late 1970s who also influenced policy. In 1987 the Greens were the third most important party in the federal Parliament.

Some 7,000 km of planned motorways were cancelled because of their negative effects on 'nature and landscape', and the priority for road construction switched to bypasses for small and medium sized cities. The importance of public transport has been extended from its promotion in urban areas to its retention in rural areas. Because of financial and budget restrictions, it now becomes important to evaluate alternative investments more systematically, principally through cost benefit analysis. For many public transport schemes it was also necessary to carry out a wide social analysis of benefits, particularly in rural areas and where there were wider development benefits. There was an undertaking to subsidize public transport for its social function. The subsidy to local passenger transport increased from 4 billion DM (1970) to 11.8 billion DM (1980), with the federal contribution falling from 68 to 58 per cent.

The *Länder* play a key role in transport planning as funds for local road construction and local public transport support are distributed through them. Under the *Gemeindeverkehrsfinanzierungsgesetz* (Community Transport Financing Act) (1975) projects can be funded by up to 60 per cent from the federal government. Over the period 1967–1981 some 23 billion DM were spent on S-Bahn, U-Bahn and Stadtbahn schemes in Germany, with the federal government funding 56 per cent. The remaining 44 per cent was split evenly between the *Länder* and local authorities (Girnau, 1983).

The concern over increasing levels of public transport subsidy has resulted in organizational changes and the creation of *Verkehrsverbunde* (Transport Associations) in the major German cities. These partnerships between transport operators have allowed the pooling of revenues and resources, the effective coordination of public transport services, the common marketing of services, and the integration of fares structures. Although the ultimate control remains with the separate operators and the public authorities to which they are responsible, there is considerable cooperation between operators established as a series of direct commercial agreements. All local and national operators are included and the partnership contract specifies the duties of the different partners including their rights and obligations. The revenue is pooled and distributed according to patronage and the costs of meeting the *Verkehrsverbunde* specifications. In Hamburg the HVV (the Hamburg Transport Association) replaced eight agencies responsible for the provision of transport (1965), and similar *Verkehrsverbunde* were subsequently established in Munich (1972), Frankfurt (1975), Stuttgart (1978) and the Rhine-Ruhr (1980). Such an organizational change allowed greater quality of service to be accompanied by greater efficiency, with cost recovery levels from fares being improved to 65 per cent. A greater weight was placed on coordination of services than in deregulated Britain, yet it seems that efficiency was also improved and patronage levels maintained despite the size and complexity of the *Verkehrsverbunde*.

Since 1980, comprehensive transport plans have been produced for cities, for states and for the federal state. These plans have formed the basis of the coordinated transport investment programmes. The aim of the plans is long-term forecasts of demand with the objective of matching up supply of infrastructure for road traffic and public transport. The methods used have not been general land-use transport studies but more focused investigations, such as the effects of policies to switch demand from car to public transport or to non-motorized transport. Analysis considered information and marketing strategies as well as infrastructure related policies. Efforts have also been directed at developing multicriteria evaluation methods to assess transport investments, including environmental pollution, land use, traffic and financing aspects (Wermuth *et al.*, 1990).

The third dimension of German policy has been the concept of *Verkehrsberuhigung* (Traffic Calming) in city centres and residential areas. By the mid-1980s, it seems that most German towns with over 50,000 inhabitants had a pedestrianized area (Hass-Klau, 1990). It was realized that the car and the city were incompatible, and to maintain vitality of the city centre it was necessary to create 'liveable' streets (*Wohnstrassen*). Close parallels can be drawn with the Dutch *Woonerven*.

The planners had been instrumental in widening the debate on urban renewal from the concern over poor housing conditions and a lack of opportunity to consider the urban environment and the quality of life. The initiative had come from the Federal Ministry of Regional Planning, Housing and Urban Development (*Bundersministerium für Raumordnung, Bauwesen und Städtebau*) and was in effect a covert criticism of the Federal Ministry of Transport (*Bundersministerium für Verkehr*) which was still advocating maximum accessibility for cars. Much of the research has been carried out by urban planners, geographers and architects, not the traffic engineers, and the case was based on safety as well as the environmental arguments.

Since the uneasy early alliance, traffic calming has become a broad based transport and planning tool which integrates all forms of transport with urban development. The 30 km/h speed limit in residential areas was combined with good design to ensure correct priorities were allocated to people and vehicles. More fundamentally, calming may have altered public attitudes to the car. Its dominance and attractiveness is not being questioned, but people and planners are no longer blinded by its limitations and the basic incompatibility between pedestrians and cars.

Germany may be moving towards a new transport policy (Zopel, 1991). Traditional transport policy and planning aims to avoid or remove bottlenecks by providing new roads and parking facilities. Provision for the car encourages greater car use and aggravates the problem it is trying to solve. What is needed is an environmental transport policy with public transport being provided as an attractive and pleasant alternative. Transport wastage is leading to a traffic and environmental crisis. A better quality of life and a better environment can only be achieved through an acceptable level of mobility and a high degree of efficiency. Although this statement is based on a political comment by a Social Democratic member of the Bundestag, it does suggest that Germany may be moving beyond the car.

## COMMENT

The Germans seem to have been more active than the British in building new roads around and between cities, in investing in and subsidizing public transport, and in calming traffic in city centres and residential areas. Rigorous change has been implemented on three fronts simultaneously, whilst in Britain and in France a more cautious approach has been adopted. The federal structure allows central government to set the framework for planning and transport, and to control strategic policy on public transport. The decentralized *Länder* have responsibility both for land-use controls and for public transport. In France decentralization of power has also given the regions greater autonomy, but in Britain the reverse has occurred with increased centralization and control over local authorities, through limitations on the ability to raise revenue locally and through cash limits imposed by central government on local expenditure. In Germany, the concept of subsidiarity is much more commonly used than in Britain. Decisions are made at the local or state levels and each territorial unit is financially responsible for its fixed transport assets, including public transport. But the federal government still has a very strong influence in local investment decisions through project linked funds. The 'golden rein' still operates in local decision-making.

Changes in approach have been brought about by powerful pressure groups. These include the transport enterprises, especially publicly owned firms, syndicates of trucking firms, trade unions, and automobile clubs. There are nine million members of the automobile club and some 38 million drivers in Germany (the old West Germany). Conversely, the rise of the Green party and the environmental movements in the 1970s and 1980s has also been influential in developing alternative strategies for urban areas, including traffic calming and heavy public transport investment. These environmental groups have a secure political base and a broad based programme from which to operate. In Britain, pressure groups tend to be more restricted in their actions, concentrating on particular issues. Their political base is weak and the electoral system does not allow them full representation in politics at the central or local levels.

The strong engineering tradition in Germany in transport analysis is still apparent, and the conceptual framework established in the 1960s is dominant. Rather than involve social scientists in analysis (as in Britain) or to have a different tradition of social analysis (as in France), the German engineers have themselves diversified to embrace disaggregate methods and the situational approach. Social scientists have tended to concentrate on small-scale empirical surveys which are time consuming, expensive and difficult to make generalizations from. There is a need for a coordinated series of investigations where more general data are supplemented by a series of in depth studies which would allow comparison and a common reference base to be set up – KONTIV is one example of such an approach. These data also indicate that although distances travelled in Germany have increased, the overall travel time of people has remained relatively constant (Brög, 1992).

## Transport Planning in France

Transport planning in France falls into two distinct phases, the first of which was concerned with road investment to catch up on construction after a long period of stagnation under rail-oriented state planning in the 1950s. The second, started in 1973, marked the rehabilitation of public transport. National planning in France had traditionally been top down with systematic data collection, forecasting and monitoring with the national institute (INSEE:*Institut National de la Statistique et des Etudes Economiques*) being the focal point of this activity (Benwell, 1980). There was considerable political power in the car industry and in all the National Plans produced in the 1960s and 1970s increased car output was anticipated. The Fifth Plan (1966–1970) demonstrated a 'Buchanan like concern with the need to improve highways and to adapt the town to the car.' The thinking was that the car was the only viable means of travel and huge urban investment plans were approved with the construction of inner ring roads.

It was at the end of the 1960s that financial limitations were imposed and other factors such as the quality of life and the indispensability of public transport became more prominent. The first bus priority schemes had been introduced in Paris and Marseilles in 1964, and political pressure was imposed by public transport operators for increased government investment. The *Loi d'Orientation Foncière* (Land Guidelines Act) (1967) permitted the grouping of adjacent local authorities for planning purposes so that informal strategic plans could be drawn up. Perhaps the most important single factor which triggered many changes in attitude and policy was the student riots of 1968.

In 1970 the *Colloque de Tours* called for integrated multi-modal transport and a rebalance of investment towards urban public transport. This meeting was the first political initiative of the public transport lobby since World War 2. The professional focus at the *Ponts et Chaussées*, the main training centre for French traffic engineers, was still oriented towards highway construction and traffic planning, but with the enlarging of the profession and political pressure decisions were reversed. A few critical urban motorway proposals were abandoned such as the left bank of the Seine expressway through Paris, and large-scale public transport investment took place in Paris with the first *Réseau Express Régionale* – RER line, the complete refurbishment of the Métro and the bus fleet. In addition the decision was made to construct other RER lines and métros and light rail networks elsewhere. The Marseille métro was opened in 1977 and the Lyon métro in 1978. Public transport professionals were able to protect their programmes through contracts between agencies (*Régie Autonome des Transports Parisiennes* – RATP and *Société Nationale des Chemins de Fer* – SNCF) and devolved planning agencies replaced direct government control and supervision.

The 1970s was the decade of public transport with the Sixth Plan (1971–75) and the Seventh Plan (1976–1980) both marking heavy investment in the public transport network and the balancing of financial objectives by social ones. By the end of the decade energy savings were also considered important as France imports over 75 per cent of her energy. Local transport planning was renewed in 1973 (*Dossier de Transport*) with a 10-year strategy statement and a five-year programme with a priced tactical programme, and consisted of a set of long-range plans heavily oriented towards road and motorway construction. Each town with a population over 80,000 was to prepare a plan. Local authorities took over ownership of the public transport network and a local transport tax paid by employers, *Le Versement Transport*, was introduced in 1971 in Paris. In 1973 it was extended to cities over 300,000 and in 1974 to towns over 100,000 (table 5.1). By 1988 *Le Versement* was funding 30 per cent of the costs of public transport (users 45 per cent and local government 25 per cent). The *Carte Orange* was introduced in Paris and there was greater user representation in decisions. 'Urban transport policies in the 1970s can be analysed as local takeovers of national procedures' (Lassave and Offner, 1989). This municipal takeover of urban transport resulted in a significant improvement in the quality of public transport, the protection of town centres against the car, and an attempt at the evaluation of non-motorized transport and the needs of particular social groups.

The separate plans were produced for each town of more than 10,000 inhabitants. A long-term land-use plan (*Schéma Directeur d'Aménagement et d'Urbanisme*), which had been in use since the late 1960s, looked 30–40 years ahead and identified broad land use zonings. By 1980, some 370 SDAU had been started, covering 26 per cent of the area and 70 per cent of the population. As with the comparable Structure Plans in Britain, progress was very slow even though they were more modest in scale. A medium-term land-use plan (*Plan d'Occupation du Sol*) zoned all land including conservation areas, urban development land, and locations with special controls. By 1980 some 11,400 POS had been designated (Simpson, 1987). Transport studies were complementary to these zoning plans and can be seen at five levels (Ferreira, 1981).

1.   Long-term strategic studies interacting with the SDAU and financed by central government. The main purpose of these studies is the identification of the major corridors for transport in cities and towns (over 20,000).

2.   Medium-term studies (20–25 years) to determine the exact land requirements for a specific facility. These studies must take the medium-term land-use plans (POS) into account.

3.   Investment programming studies. These are five-year plans containing a detailed priority list of transport improvements. Such studies are carried out at the city level and updated annually (*Dossier de Transport*).

4.   Detailed costing studies of specific projects.

5.   Detailed design and documentation for specific projects.

Since the global and national economic transformations of the 1970s, the end of long-term simple mechanistic forecasting methods was becoming possible. The social nature of communications needs was now recognized as the basis of physical movement. However, traditional methods are difficult to replace completely and although it was now possible to take a broader social-based approach to analysis, there were still aspects of these methods in evidence. The change was not clear cut. The 1980s marked this change with a decentralization of decision-making. France's modernization of local government came very late (1982–85) and the 36,000 communes have still been retained. Local authorities already had powers for public transport, traffic, parking and town planning, and now land use and traffic management were added.

The *Loi d'Orientation des Transports Interieurs* (LOTI) (1982) was directed towards 'the right to transport for all', as all users should now have the right to travel and the freedom of choice of mode with reasonable access and at a reasonable cost. Priority for public transport was to be balanced by complementarity with other modes and by fair competition. Urban areas were encouraged to prepare urban travel plans (*Plan de Deplacements Urbains* – PDU) with a statement of policy to cover traffic systems and priorities for all modes of transport. In theory, part of the thinking here was to reverse the trend towards urban sprawl, to give priority to public transport to increase its efficiency, and to reduce the dependence on the car. The PDU also introduced a public inquiry procedure to open up the debate between the population and the professionals and politicians, but also to get their agreement and support for proposed policies.

Within LOTI (1982) there was also a requirement for an infrastructure master plan to cover an evaluation of alternatives including social and economic factors, and this document should be made public before the adoption of a project. The *Schèmes Directeurs d'Infrastructures* are prepared by the state in consultation with the regions and by the municipalities (Simpson, 1987).

The *Versement Transport* has also been extended to all towns with populations over 30,000 (1982). As with the larger settlements, the implementation of the tax is optional, but most local authorities have taken the opportunity to introduce it at the maximum permissible levels. All employers with more than nine wage earners must pay the tax, and the levels are calculated on the basis of salary ceilings. In 1988 the level was 9,950 FF/month and the application of fixed percentage rates (table 5.1). In Paris most of the revenue is used to pay for tariff reductions given to employees (*Carte Orange*) with little contribution to depreciations and nothing to finance new infrastructure. Receipts in Paris are now declining in real terms as decentralization takes place. If additional revenue is sought, the only option would be to raise the fixed rate again. Elsewhere, of the 125 possible locations, 104 had introduced the *Versement Transport* (1986) with some 60 per cent of revenues being allocated to revenue support and the remainder being used for capital investment and the repayment of interest on loans.

The *Versement Transport* allows some continuity to public transport funding as the operators know the approximate levels which will be raised each year. The salary ceiling and the level of the tax may be changed, but that continuity allows longer-term decisions to be made. In the USA and Germany the tax on petrol with an allocation to public transport is an alternative way to fund loss-making bus and rail services, and to encourage investment. In Britain there is no such continuity and public transport operators are dependent on fare revenues, tight cash limits on new investment, and the annual budgets of central and local authorities. Long- or medium-term investment decisions are difficult to make under such conditions of uncertainty. Privatization of bus services means that operators can now raise capital on the commercial markets, but even here payback periods are lengthy and the returns do not seem attractive. Much of the explanation for the poor quality of the British transport infrastructure, particularly in public transport, can be explained by the failure of successive governments to establish a stable financial basis for long-term investment, either through an employment tax or through a petrol tax or even through facilities to borrow capital at attractive interest rates as happens in France.

In the 1980s there was also a switch from the focus on urban to interurban applications of formal evaluation and modelling. Urban evaluation has become very pragmatic for small projects with a strong qualitative orientation,

TABLE 5.1. History of the *Versement Transport*.

| 1971 | Inner Paris | 1.7 per cent |
| 1973 | Cities over 300,000 | 1–1.5 per cent |
| 1974 | Cities over 100,000 | 1–1.5 per cent |
| 1975 | All Paris Transport Region | 1.2 per cent |
| 1982 | Towns over 30,000 | 0.5 per cent |

and it can be seen (as in Britain) very much as a bidding process for resources. Large projects such as the *Train à Grand Vitesse* (TGV), the Channel Tunnel and the RER lines require more classical and formal models.

The TGV Sud-Est line between Paris and Lyon (427 km) was opened in 1981–83 with a further extension to Grenoble in 1985 to carry passengers at speeds of 270 km/h along new specially constructed track. The cost was FF 8.5 billion (1985) with a further FF 5.7 billion (1985) for rolling stock. The journey time between Paris and Lyon was cut by 1 hour 50 minutes to 2 hours, and traffic increased by 45 per cent to 14.3 million (1983–84) and a further 10 per cent to 15.8 million (1985). Formal economic evaluation calculated an economic rate of return of 15 per cent and a social utility rate of 30 per cent. The social utility broadens the evaluation to include the public interest and gives a value about twice that of the financial viability. These factors include safety and the environment as well as structural changes which may be induced in regional planning and development. The capital was raised through borrowing by SNCF after the project had been approved centrally. There is also an inquiry prior to the formal approval by the state so that individuals can make representations on the basis of a report presenting the project and including an impact study. SNCF has excellent credit standings on the international market because of government guarantees (Gerardin, 1989*a*, *b*).

By 1984 the TGV Sud-Est generated FF 3 billion in revenue of which 35 per cent covered operating costs and there was a return of FF 400 million after debt servicing and depreciation costs. The levels of traffic exceeded predictions with the following proportions coming from other modes or being generated (Bonnafous, 1987):

33 per cent from aeroplanes

18 per cent from road

49 per cent from induced traffic (including rail)

Apart from economic and financial appraisals, the TGV is also subject to environmental and regional impact analyses. For the services to be profitable the links had to be from Paris and this has meant that links between regional centres have not been improved. Paris has become the national hub. A series of surveys have been carried out along the Paris-Lyon corridor and these indicate growth in tourism generated by the TGV as well as growth in high level service industries (Bonnafous, 1987). However, the TGV has had little effect on the places along the route, only a powerful one at both ends. The Paris-Lyon link broke the psychological barrier as the return journey could be made in a day. The development impacts have been selective and seem to have reinforced existing trends. The TGV is a communications device which has changed the face of transport, but it alone does not change business decisions or the distribution of activities.

Subsequent decisions to build a TGV network in France were predicated on the success of the TGV Sud-Est. The French government adopted (1989) a master plan for high speed rail links (table 5.2) and other links are being proposed with lower internal rates of return and social utility. With the extension of the TGV line to the airport at Satolas (NE Lyon) for the 1992 Winter Olympics (Albertville), a choice from four routes was made. These options were compared by three experts appointed by the Ministry of Transport and they recommended the rural route. Counts were made of buildings between 40 and 200 metres from the axis and every effort was made to place the new railway next to the existing Lyon-Geneva motorway to minimize the environmental costs. The final decision of the Minister can only be challenged by instituting action in administrative courts and this may result in more protection without altering the layout, for example by the more extensive use of tunnels. Once the decision has been made the time between decision to proceed and opening of the project is short (4–5 years).

In France there is a broader based approach to evaluation of the TGV network which balances the economic and financial costs and benefits against regional development and environmental impacts. Financing is a complex issue as regional and European contributions have to be balanced against the part that SNCF could (or should) pay for. In addition, the assessment must ensure that new links do not undermine the viability of the network as a whole as the potential for traffic and revenue generation is considerable. The public inquiry stage is limited to debates on actual routes and the use of expert advisors is standard practice in both road and rail investment decisions.

The difference between the British and French system is that in Britain proposals are produced for local reaction whilst in France proposals are agreed and then discussed locally. It seems that France has developed a set of

TABLE 5.2. The development of the TGV network in France.

| Title of TGV | Route | Length (km) | Opened | Cost (1985) | IRR (%) | SU (%) |
|---|---|---|---|---|---|---|
| Sud–Est | Paris–Lyon | 427 | 1981–3 | 13.2 B | 15 | 30 |
| Atlantique | Paris–Bordeaux | 580 | 1989–90 | 16.5 B | 12 | 20 |
| Est | Paris–East France and Germany | - | - | 23.0 B | 3.5 | 9 |
| Nord | Paris–Brussels, London, Cologne, Amsterdam | 300 | 1993 | 28.5 B (1987) | 7 | 11 |

*Notes:*
IRR is the Internal Rate of Return;
SU is the Social Utility;
Costs are in billions of French Francs.

procedures and the organizational and financial framework necessary for a new generation of rail investment to produce a Europe-wide high speed rail network. In 1988 the Community of European Railways, which includes the 12 EC members plus Austria and Switzerland, proposed a high speed network based on existing short haul air routes and the likelihood of a quadrupling of traffic over the next 30 years. This new network would consist of 12,300 km of upgraded and new rail by 1995 and a final length of 30,000 km at a cost of 100 billion Ecu (£70 billion).

COMMENT

Transport planning analysis in France has had less emphasis on formal mathematical models and a much greater focus on simple empirical methods. Some modelling is required for the National Plans (carried out by INSEE), for the SDAU Plans, the more detailed *Dossier de Transport* and for all large projects. The urban division of SETRA (*Service D'Etudes Techniques des Routes et Autoroutes*) has compiled a set of manuals for strategic transport planning where accessibility has to be improved without any loss of quality of life. However, as early as 1973, questions were raised about massive data collection exercises and there were doubts in some circles about the relevance of large-scale models to the development of transport policy (Dupuy, 1978). France has never fully believed in the systems approach to transport planning, and has placed a greater emphasis on demand oriented analysis.

Trip generation and attraction rates were usually empirically derived and used in a gravity model to determine trip distributions. For example, the gravity model was used for the evaluation of the TGV Sud-Est railway, based on the generalized costs of travel between Paris and Lyon. Modal split was carried out on the basis of past trends, the options being tested with empirically derived utilization curves. Studies of central area parking demand and supply were carried out and the highway network examined to see whether capacity was exceeded. Cost estimates were made of the changes required to maintain the level of environmental quality and accessibility. There was feedback here to the modal split analysis. Assignment was carried out by the simple procedure of allocation of trips between each zone to the network by mode. There was a considerable amount of judgement used by the analyst in allocating traffic between two similar routes. For public transport, the network was defined to be consistent with the SDAU and the requirement to maintain good central area accessibility. Options were developed on the basis of demand using average figures per head of population, on accessibility as measured by trip times from home to work, and on urban structure.

In all these procedures there is a close link between land use and transport. Transport plans, both long term and medium term, form an integral part of urban planning and are incorporated into the land-use plans (SDAU and

POS). In preparing the SDAU, estimates are made of the amount and distribution of economic activity as well as the distribution of population. Other features such as public buildings, parks and recreational areas are identified. The transport studies estimate future peak hour traffic volumes and central area parking requirements. Future infrastructure needs are then estimated under a range of alternatives from heavy transit to private transport. A dialogue between politicians, planners, engineers and economists then follows to arrive at the best possible compromise solution.

This summary, taken from Ferreira (1981), leads to the conclusion that transport planning links with urban planning are stronger in France than in Britain, but that this consistency is at the expense of internal coherence of the transport planning process itself. Rather than having two separate analysis processes with little interface between them, the French have followed a strong planning philosophy with transport analysis being brought in to help identify problems and to test possible solutions. This is one attempt at a better integration of land-use and transport analysis.

This fundamental difference in approach can be explained by a series of related factors. Substantial transport research in France has been based in non-technical research traditions of sociology. As Benwell (1980) comments the French have lacked 'the impetus of the background questioning and assumptions which have led many USA and UK researchers from mathematical and engineering backgrounds into econometric and attitudinal studies. Many French researchers have extended their interests into a transport context from a background in other branches of the social sciences.' The modelling problem-solving orientation of much British and American research is replaced by a more consciously theoretic conceptual orientation of linkages between transport and land-use dynamics.

This different theoretical background allows many different perspectives on the behaviour of individuals, the role of cultural factors, and the organization of society to be assessed in the travel decision. At the macro level it also allows explanations of major policy changes. For example, it is generally recognized that fundamental changes took place in 1970–72 on investment policies for the Paris Métro as a result of large-scale user protests and pressures exerted by the Left. It also allows analysis of perceived irrationality of some public transport policies which were adopted. The RER A line was built to make the La Défense business complex viable. The needs of manual workers peripheralized to the outer suburbs by the increasingly homogeneous gentrification of inner Paris were not considered (Dunleavy and Duncan, 1989).

The focus was often on small-scale intensive studies once it was realized that sophisticated forecasting methods were not needed. Studies usually only took about six months to complete and there were often non-experts on all technical committees which meant that the techniques had to be simple and transparent. The models developed were not disaggregate or behavioural, yet the results seemed to be no less accurate than the more sophisticated approaches. A study which compared observed with modelled values from 17 home interview surveys in France found that errors ranged from 7–15 per cent of the various trip generation categories, and that for trip distribution the errors were 25 per cent. These figures compare favourably with those found in Britain (Evans and Mackinder, 1980).

## Transport Planning in the Netherlands

It is in the Netherlands that transport planning seems to have maintained its position as central to environmental and national planning policy. Rather than moving away from planning as an approach to problem solving, the Dutch have recently (1989) produced a second National Transport Structure Plan (*Structuurschema Verkeer en Vervoer II – SVV-II*) to cover both passenger and freight transport to 2010. A series of scenario objectives has been set so that regular monitoring can take place and modifications can be made in the more detailed implementation document, the Infrastructure and Transport Programme. Specific policies fall into four categories, each of which is of equal importance:

*Category 1: Improving Accessibility* – the provision of a high grade network of national and international roads, railways and waterways serving the goal of *Nederland Distributieland*. As national income rises, the transport industry's share can be held at not less than 7 per cent. Transport proposals must be linked in with patterns of development proposed in the Fourth Report on Physical Planning.

*Category 2: Managing Mobility* – measures to reduce the use of the car by discouraging avoidable car use and ensuring that attractive alternatives exist. The target for 2010 is for an increase of 30 per cent in peak hour car use (as compared with 1986) rather than the predicted growth of 70 per cent if no action is taken. Use of public transport at the peak will double and long distance rail will increase by 50 per cent.

*Category 3: Improving Environmental Quality* – emissions by motor vehicles will be significantly reduced and there will be no increase in noise nuisance (1986–2010). Traffic will be concentrated on through routes, road safety will be improved (50 per cent fewer deaths), and the growth in road freight traffic will be reduced. Interestingly, only emissions of nitrogen oxides and unburned hydrocarbons are covered in this category. Other emissions such as carbon dioxide were covered in the National Environmental Plan (NMP+).

*Category 4: Support Measures* – provides the means to achieve the first three objectives through the reforming of transport finance and the applications of the user pays principle; the development of better coordination in transport policy, through better enforcement, through consistent cooperation between transport agencies and through positive exploitation of market opportunities.

As Gwilliam (1990) comments, the concept of a comprehensive transport plan is one which commends widespread support in the Netherlands, but it is clear that it contains substantial weaknesses. He cites inconsistencies in objectives, the lack of adequate definition of the capability of instruments, the separation of structure planning and financial planning, and the political vulnerability of end state planning. However, it does act as a focus and framework within which to structure more detailed analysis and policy-making, and it is consistent with the targets set in the National Environmental Plan (*Nationaal Milieubeleidsplan*) and its extended version (NMP+).

Within this national framework of transport, environmental and land-use plans, significant amounts of detailed research are carried out often using disaggregate analysis methods and novel forms of data collection (for example panel surveys and the mobility scanner). The strength of strategic planning in Netherlands seems to be at odds with other trends which have taken place in policy-making including less government intervention, the desire to reduce public expenditure and the increasing levels of international integration. Instead of moving towards allowing the market to determine levels of demand and prices, policy in the Netherlands seems to be questioning the broader issues of location, the increases in mobility and the negative external effects of the car. Financial support is being used to fund new infrastructure rather than subsidizing existing services, and new approaches to public management are being sought. It seems that the Netherlands has reached a decision point where all the negative impacts of the car and high levels of mobility have become apparent within a small country – congestion and a lack of capacity, heavy subsidy in public transport, urban deconcentration, its position as a major distribution centre for northern Europe, environmental pollution, the role of telematics and broad band telecommunications, and high speed rail connections. The recent changes in policy towards strategic planning reflect both the severity of the problems and the means by which they can be addressed through a mixture of professional pragmatism and scientific research (Priemus and Nijkamp, 1992).

Ambitious targets have been set to reduce levels of transport pollutants. Nitrogen oxides and hydrocarbons will be 25 per cent of their 1986 levels in 2010, and carbon dioxide emissions will be set at 90 per cent of the 1986 levels. Road deaths and serious injuries will also be halved over the 25-year period (Rietveld, 1993). The Dutch government gives priority to reducing acid rain rather than global warming (carbon dioxide) and the consumption of fossil fuels (oil). Targets have also been set for maximum acceptable congestion probabilities of 2 per cent on the main corridors and 5 per cent on other parts of the road network.

To achieve these targets, strong political action is required as the expected growth in traffic over the review period (1986–2010) will be 70 per cent. Fiscal measures being considered include road pricing and a real increase of 50 per cent in fuel prices which in turn will affect the fuel efficiency of the vehicle stock. Tax concessions available at present on long-distance commuting by car (over 30 km) would also be removed. Industries and offices are encouraged to make use of shuttle services with private sector coaches. A fourth possibility is a doubling of real parking tariffs (Rietveld, 1993). Increases in the length of the highway network will be limited, but the capacity will be increased by 30 per cent through additional lanes, and the use of telematics will give an extra 15 per cent capacity per lane. Priority will also be given to public transport, car pooling, and bicycle use (bicycles still account for 28 per cent of all trips in the Netherlands). Physical measures include strict restrictions on parking spaces for firms and the requirement for firms to develop trip reduction plans as in the USA (see page 100). Transport regions are being set up to coordinate and implement the regional transport plans.

COMMENT

Although these proposals set new standards for environmental and transport policy objectives, achievement of the targets presents real problems as links have to be made with the general economic development of the country. Growth since 1986 has followed the high growth scenario, and it is only more recently that modest growth has been achieved. Much faith has been placed in the SVV-II and NMP+ strategic models and it is unclear whether the policy measures will result in the expected reductions in the use of resources and levels of emissions. The actual responses of individuals may be to accept higher prices so that they can continue to use their cars. It also depends on the government implementing policies which may be unpopular. Road pricing in the Randstad is one of the main instruments for reducing levels of congestion and use of resources in transport, but it is unlikely to be introduced as strong resistance has been roused by various interest groups and political parties. Road pricing has been rejected by the Second Chamber in Parliament.

> If road pricing would only be located in Randstad, it would create a negative location factor for that part of the country and would increase the trend to disperse industry and housing. Incidently, the social costs of the Randstad are also the highest in the country, which could conceivably justify higher variable costs for automobile use. (Priemus and Nijkamp, 1992, p. 185).

Toll pricing at key points on the network is the alternative most likely to succeed, particularly if private finance is used to build five new tunnels – the

Wijk and the second Coen, near Amsterdam; and the De Noord, the second Benelux and the Blankenburg, near Rotterdam. Other forms of user charge, such as an increase in fuel tax, are likely to be introduced at the European level as a carbon tax.

It seems that although much analysis has been carried out and a range of major policy measures has been proposed and agreed in principle, the reality is more problematical, particularly if it is seen to be incompatible with the central government's intention to reduce its substantial budgetary deficit. The introduction of such measures, particularly in a recession would make the Dutch economy less competitive. In all countries, when the economic imperative is overriding, all significant policies that might actually reduce levels of mobility growth and lead to environmental improvements become of secondary importance.

## European Approaches to Evaluation in Transport

The differences in approaches to transport planning are crystallized in the important issue of evaluation, which forms a crucial part of project assessment. In Germany, the Federal Transport Plan (*Bundesverkehrwegeplan*) requires a benefit cost study for all projects over 500,000 DM. In this evaluation the alternatives are ranked by three criteria: traffic and congestion effects, regional policy objectives, and other effects such as the elimination of accident blackspots and their international importance. Where possible all costs and benefits are presented in monetary values and results are presented as benefit cost ratios and as net present values. The discount rate recommended is not the market rate, but a rate of 3 per cent which is chosen on the basis of the expected national economic growth rate, in real terms.

Demand forecasting in Germany takes a range of values and uncertainty is covered by using less optimistic demand forecasts. Regional impact assessments on structurally weak and peripheral areas, together with environmental compatibility assessments, are all part of the standard procedures. In response to the EC Directive on Environmental Impact Statements, an estimation of ecological risks along the proposed route is now investigated, and guidelines for risk estimation have been produced. The environmental compatibility study defines the broad area for investigation, the planning of route alignments, and a detailed explanation of the advantages and disadvantages of each.

Road evaluation in France switched from the primary concern of returns on capital towards a broader based multicriteria approach. The Directive for Road Transport Evaluation (1986) uses 10 categories for evaluation (table 5.3) of both monetary and non-monetary factors which are then brought together in a series of matrices that rank the options and demonstrate their positive and negative characteristics. The flexibility introduced by such a procedure seems to give the impression that decisions are taken without recourse to formal procedures. Evaluation is just one factor in decision-making and facilitates discussion between the major parties concerned. As with Germany's approach to evaluation, public hearings seem to have only an indirect influence on the planning of roads and other projects as there is no formal presentation of the results of the evaluation to the public. Often the route is decided and individuals have to resort to the courts to change that decision. Even then the decision is unlikely to be reversed, but minor changes including environmental improvements may be made. This means that the time taken from project inception to completion can be considerably reduced. The lead time is often only five years when in Britain it takes over 15 years on average to complete all the procedures (see pages 57–60).

In both France and Britain the importance of development objectives is considerable. In France, DATAR (*Délégation à l'Aménagement du Territoire et à l'Action Régionale*) argued the case that the original Master Plan for Roads (1960) favoured the prosperous central areas as it was based on the predicted continuation of existing trends which in turn suggested that the greatest economic rates of return would take place in these core areas.

The DATAR plan used criteria for national and regional development to promote greater investment in peripheral regions. However, the government was also concerned about restricting the levels of public funding of the motorway systems, and were concerned with encouraging private concessions for toll road construction. This problem has been augmented by the move towards decentralization of power in France. The regions can now use their powers, their considerable financial resources and political will to favour free

TABLE 5.3. Criteria used in the evaluation of road proposals in France.

1. *Regional and Local Development* – ease of movement, indirect employment effects, changes in attractiveness of an area and the balance between developed and less developed areas
2. *Safety* – reductions in numbers of deaths and severe injuries
3. *Environment and Quality of Life* – effects on natural resources and ecosystems, human activities including urban planning and access, and quality of life
4. *Minimization of Severe Problems* – alleviation of particular problems such as blackspots, congestion and risks from natural phenomena
5. *Impacts on Other Transport Modes* – switches between modes including gains and losses of revenue
6. *Direct Effects on Employment* – jobs created in the construction and maintenance of the project
7. *Energy and Balance of Payments* – net change in energy consumption, tourism and taxation
8. *Public Accounts and Accounts of Concessionaires* – net change in expenditure and receipts for public expenditure and the ability for private operators to recoup costs
9. *Cost Benefit Assessment* – all possible factors are included to give a net present value and first year rate of return

motorways or roads of an equivalent quality. But the central government now favours public transport, especially rail, and the private funding of road schemes (Grandjean and Henry, 1984).

This dilemma is best illustrated in the weights attached to costs in cost benefit analysis if public funds are being used (*Coefficients de Restriction des Crédits*). The weights (1.5 or 2.0) are only used for public funds to reflect their scarcity and these values obviously reduce the attractiveness of any possible road investment. Similarly, a discomfort penalty is used on the benefits (*Malus d'inconfort*) so that if the road is not a motorway, then a value is added to the evaluation:

2 centimes/km/car for dual carriageway roads with graded intersections;
8 centimes/km/car for other dual carriageway roads;
14 centimes/km/car for single carriageway roads.

Naturally, these discomfort penalties will favour the construction of motorways and these penalties have caused more debate in France than the justification for the values of time used in evaluation.

Table 5.4 summarizes the approaches to road and public transport evaluation used in Britain, Germany and France. The formal use of cost benefit analysis which keeps the power in the hands of the planners is more likely to be found in Britain and Germany. Multi-criteria analysis gives the decision-makers more power and responsibilities to weight goals and other criteria, and this procedure is more likely to be used in France for road evaluation. Other forms of evaluation, such as the use of participatory methods, which involve the use of unsophisticated techniques and extensive debate with the people affected, seem to have been ignored. In Britain, the financial and economic factors dominate evaluation, but in the other two European countries regional policy objectives and environmental factors add to the financial and economic criteria.

## Conclusions

Transport planning analysis over the last 30 years in all countries mentioned here has been dominated by an engineering approach in which the quantitative values (for example capacity and network expansion) are considered more important than the qualitative factors (for example safety, distributional issues and externalities). Demand has been uncritically accepted as given and the job of the transport planner was to accommodate that demand, principally through permitting the increased use of the car. Trend extrapolation and demand forecasting methods were used to estimate future growth in demand. Again, the aim of the analysis was to test alternative strategies to meet the expected growth in demand.

At the same time the context within which transport was operating has itself

TABLE 5.4. Differences in evaluation procedures.

| Mode of Transport | United Kingdom | Germany | France |
|---|---|---|---|
| Roads | 1. Cost benefit analysis 2. Appraisal framework 3. Manual of Environmental Appraisal | 1. Strategic evaluation 2. Cost benefit analysis 3. Separate environmental evaluation 4. Regional policy objectives important | 1. Multi-criteria analysis 2. Ranking of projects 3. Limited use of cost benefit analysis 4. Development effects very significant |
| | *Comprehensive and Complicated* | *Comprehensive and Complicated* | *Transparent and Simple* |
| Public Transport | 1. Financial appraisal 2. Some limited cost benefit analysis | 1. Cost benefit analysis 2. Utility measurements for non-monetary factors (land take, noise, pollution, energy, comfort and quality) 3. Verbal assessments for intangibles | 1. Limited formal measures 2. Internal rates of return and social utility 3. Regional development impacts 4. Environmental impact studies 5. Use of expert views 6. Revenue cost ratios |
| | *Capital Costs should be Recouped from the Users of the Facility* | *Eligibility for Grants Depends on use of Procedures* | *Public Transport seen as a Social Benefit* |

changed, with demographic and other structural changes in society, with increases in leisure time, with changes in lifestyles and the growth in levels of affluence, with shifts in the labour market, with technological developments, and with concern over qualitative factors such as the environment. Public policy-makers in transport were also facing a series of difficult questions. There was a growing backlog in road and rail infrastructure maintenance and reinvestment; there were the insoluble problems of congestion in urban areas; there were unacceptable delays to public transport users on the network and at termini; there were threats to the viability of some public transport services; there was an awareness of incompatibilities between the switch to the car and environmental objectives; and there was a need to accommodate expensive advanced technologies (based on logistics management) in land-use decisions. Governments were also making an important contribution to the debate. Policy priorities changed significantly in the 1980s with a concern over regulatory reform and the opportunity to increase competition within transport, with cutbacks in public budgets available for both capital investment and revenue support. Superimposed on these changes is the growing internationalization of many transport decisions as policy initiatives were being taken outside of the national arena, principally by the EC. Yet most of the available investment funds will still be used for road investment rather than public transport investment (table 5.5), with the exception of the Netherlands.

In short, the position of transport policy as a strict regulator was being questioned for a range of reasons (Noortman, 1988). These included a lack of consistent and non-conflicting objectives; a lack of adequate and effective policy instruments; limited budget capacity to implement policy actions; inertia in transport policy caused by long-lasting bureaucratic procedures; and the lack of a suitable and efficient legal system for a creative and trend setting policy.

The underlying philosophy of the public policy approach was being questioned. The two main issues related to the public goods and the externalities arguments. The public goods argument refers to the indigenous role of transport and infrastructure in society, in which equity considerations and monopolization objectives are as important as efficiency objectives. The externalities argument concerns both the positive aspects, such as stimulating economic and regional development by improving accessibility, and the negative aspects, such as the need to reduce air pollution and noise nuisance (Gent and Nijkamp, 1989). National governments may not be rejecting the public policy arguments on transport, but they are examining ways in which they can reduce the ever increasing financial commitment to transport and other public sector activities.

Superimposed on these two issues is the changing role of government,

TABLE 5.5. Intended infrastructure investment in the 1990s.

|  | *The Netherlands* | *Germany* | *France* |
|---|---|---|---|
| Road | 700 | 2400 | 6000 |
| Rail | 400 | 1700 | 1200 |
| Waterways | 200 | 400 | 300 |
| Airports | 200 | 100 | 300 |
| Total | 1500 | 4600 | 7800 |

*Note*: Ecu millions per annum.

*Source*: Coopers and Lybrand (1990).

whether it should be interventionist or let the market establish the appropriate levels of investment and prices charged. In general, governments in the 1980s have moved to the right with reductions in public expenditure and the basic strategy of charging the user the full costs of services used. Yet some governments (for example UK and USA) seemed much more willing to impose these policies of the new right than others (for example BRD and France). It is only in the Netherlands that radical change has taken place in policy and planning priorities which attempts to integrate transport planning with environmental issues, but even here there are now considerable problems with the implementation as the current economic recession continues. One of the unresolved questions of the 1980s seems to be that although it was a decade of radical political change, little fundamental rethinking in transport planning took place. There seemed to be no corresponding paradigm shift in approaches to transport planning that was in any way commensurate with the scale of the radical political changes.

*Chapter 6*

# The Limitations
# of Transport Planning

## The Theoretical Arguments

Transport planning has now emerged from a 30-year gestation period, and as maturity is reached it is now appropriate to stand back and assess the ways in which the theories and processes have responded to radically different sets of political constraints. This chapter places a theoretical framework on transport analysis over the last three decades and attempts to explain why very little change has taken place in the basic transport planning models (TPM). It presents a retrospective review of the TPM, comments on its limitations, and then identifies the main responses in terms of analytical approaches which have been developed in the 1970s and 1980s. Finally, it is argued that since 1985 a renaissance is taking place as for the first time there has been political and public concern over the way in which transport planning has been carried out. It is no longer acceptable to use the conventional methods which have serious limitations, cannot accommodate the types of changes now taking place in society, and have lost public confidence. Alternative planning structures are now required and the chapter ends by suggesting where this renaissance might lead.

Transport planning has developed in parallel with urban planning but seems to have avoided the major theoretical debates which have influenced social science thinking over the past 30 years. Two possible explanations might be that transport is immune to such debate or that the unique combination of engineers, economists and social scientists working in transport make it difficult to have such a critical review. The latter explanation may be more likely as most of the criticism has come from outside, but even then it seems to have been primarily aimed at planning in general. Transport planning seems to have escaped.

Table 6.1 attempts to place a general structure on the evolution of three strands of the debate identified by Yiftachel (1989). The view in the 1960s was that planning operated in the public interest, and that it could anticipate the needs of the public and act accordingly. This view was questioned in the 1970s from two basic viewpoints. The Marxists claimed that planning was an arm of the capitalist state and that it was merely facilitating capital accumulation and legitimation of the capitalist system. The contrary Weberian view was that the state served a multitude of societal interests which were increasingly controlled by a rational and independent bureaucracy. The corporatists argued that there was an alliance between government and industry so that each operated in the interests of each other. Since that time the theoretical arguments have become more disparate until the re-establishment of neo-classical principles and the new concepts of the company state. Here, the market is liberalized, the private sector effectively takes over, and the role of the state is reduced to a facilitating function (Banister, 1990). The interests of society as a whole are of secondary importance to the profits for the company state together with returns to their investors.

Transport has been largely immune from these theoretical debates as there has been a strong state role in terms of public expenditure, regulation and control, and in terms of public transport operations. Throughout the period of ideological debate and questioning, transport seemed to be excluded as it was perceived as a public good to which everybody had a right. The strong social policy tradition made it difficult to place any particular ideological interpretation on transport. However, this has all changed. The origins of theoretical criticisms of transport can be found in the 1960s and the 1970s with the growth in car ownership and road investment, but it really came to life in the 1980s with the radical policies of the Conservative governments.

In transport there has been a tendency for theorists to concentrate on the public/private divide and to relate this to income and the manual/non-manual divide (Dunleavy and Duncan, 1989). The pluralist intellectual position would be as protectors and sporadic interventionists, mainly as defensive interest groups protecting the current position. Although there were widespread concerns by these people over the quality of public transport, it was the urban road investment programmes of the 1970s and the public inquiry process which allowed the pluralists fully to articulate their concerns. The pluralists were often ranged against the elitists who represented the vested interests of construction companies in road building and the motoring organizations. The elitists were more interested in the exchange value than broad local use values and they also represented corporatist views (Whitt, 1982). The Marxist sociologists placed transport in the broader sphere of urban social movements by arguing that the basic conflict was between the advocates of the capitalist economy and those who saw broader social priorities in urban development (Lojkine, 1974).

It has only been in the 1980s that the theoretical debates have become dominant in transport and other sectors of public sector activity, with the re-introduction of neo-classical economics and its extension to the company state. The notions of a welfare state and public provision of services have been replaced by the market and private sector operations. Transport has been forced to adapt to this political imperative and enter into ideological and theoretical debates, which in the past it had managed to avoid. Previously, the debates on existing transport systems had revolved around different ways of organization in the public sector and whether subsidy should go to the user or the operator. There was virtually no debate about whether organizations should be in the private or the public sector, that was taken as given. Even on questions of regulation there seemed to be some agreement. The pluralists interpreted regulation as the state's willingness to intervene against vested interests of corporate business. The elitists argued that vested interests welcomed intervention as it promoted stability and allowed concentration. Even the Chicago School of economics made a similar argument as they felt industry was likely to control the regulatory agencies itself. This was because the interests of rational bureaucrats and politicians in these agencies get bound up with satisfying pressure from organized interests in their immediate environment. Industry pushes for regulation primarily as a means of limiting competition, stabilizing markets, and promoting collusion between businesses (Stigler, 1975). These different theories have now been tested in practice with extensive regulatory reform and privatization in transport.

The second and third columns of table 6.1 summarize the changes in city planning and transport over the review period, extending the likely trends into the 1990s. In both cases the cyclical nature of policy can be identified with growth, expansion and concentration. The major issues in the 1990s are technology and the environment, which will allow further dispersal of cities, and the full range of communications media will play a major role in business, leisure and other activities. The 1990s also presents one major dilemma, namely that choice between ever increasing levels of transport mobility and the political objectives of stabilizing the emission of environmental pollutants. Mobility, cities and lifestyles all have to become compatible if the objective of sustainable development is to be achieved.

As outlined in Chapters 2, 3 and 4, after the golden period of planning in the 1960s, there was a period of re-assessment and critique. It was argued that the systems approach was too ambitious and that the theory was too simple. In the 1970s it was also attacked for not being relevant to the policy issues of globalization of the world economies, the geopolitics of oil, and the industrial restructuring which was taking place. The established theories became even more irrelevant in the 1980s with the overt politicization of planning and transport. The ethos of rationality and comprehensiveness, implicit in the systems analysis approach, is impossible to maintain in such a politicized decision-making environment.

TABLE 6.1. The evolution of the debates in planning theory.

| Decade | Theory | Planning and City Development | Transport | Procedures |
|--------|--------|-------------------------------|-----------|------------|
| 1970 | Weberian analysis Corporatism Marxist analysis Pluralism | Natural expansion Containment Corridor development | Highway construction Management Public transport and subsidy | Systems analysis Incrementalism Mixed scanning |
| 1980 | Managerialism Reformist Marxism Neo-classicalism | Decentralization Renewal Consolidation Sustainability | Market dominance Gridlock | Advocacy Positive discrimination Pragmatism |
| 1990 | Company State | Dispersed cities Technological cities | Highway and rail construction | Quick response methods |

*Source*: Based on Yiftachel (1989).

Planning analysis seemed to adapt much more readily to this environment (table 6.1), with diversions away from the systems approach to incrementalism (Lindblom, 1959), mixed scanning (Etzioni, 1967), advocacy planning (Davidoff, 1965), and positive discrimination (Blowers, 1986). Planning procedures were seen as potential instruments for affecting the outcome of political processes, but all these interventions were short lived as the political constraints changed. The only stable paradigm seemed to be the pragmatic rationalization which concedes that absolute rationality and comprehension are not only impracticable but also politically impossible in the increasingly politicized environment of governmental decision-making (Yiftachel, 1989).

Although rational analytical processes should be maintained, it is argued that adaptability and flexibility are the keys to maintain relevance to the 'institutional and political situation in which the planner operates' (Faludi, 1987). This flexible pragmatism has much in common with Alexander's (1984) contingency approach to planning which attempts to synthesize research findings with normative prescriptions. But, in the short term, rational models together with evidence from empirical research and pragmatic experiences are likely to continue in use.

In transport analysis there has been no similar process of experimentation with other paradigms. The systems analysis procedures have remained supreme throughout the last 30 years with alternative approaches either being ignored or marginalized. Modification has taken place, but there has been no fundamental reassessment of the basic structure of the transport planning approach which is still consensus seeking and prescriptive. Political economists (for example Dunleavy and O'Leary, 1987) have criticized the approach and procedures used in transport planning, but with little effect. For example, it is argued that cost utility approaches to evaluation are one of the most distinctive features of technocratic policy planning in transport. These procedures bundle together contestable or implicit social and economic valuations under the guise of sophisticated planning methodology, with easily quantified or monetized aspects of issues being valued over and above the qualitative intangible considerations. This must erode the importance and scope of political debates and processes in decision-making (Self, 1975).

Even the New Right has criticized the cost utility techniques (such as cost benefit analysis) as an insidious tool used by budget-maximizing bureaucrats to justify over-supplying agency outputs at levels which would never be sustained under private market operations (Dunleavy and Duncan, 1989).

Despite these forceful criticisms of the technocratic transport planning process from both ends of the political spectrum, the procedures remain intact. There is a realization that many of the decisions are not matters of expertise but matters of opinion, of values rather than facts. In short, decisions are political but this has not made any real difference (Blowers, 1986). Transport planners have tended to remain neutral rather than play a significant role in decision-making. They have tended to scale down their goals and accept the objectives and values set by politicians (Reade, 1987). This means that they have made a positive choice to remove themselves from having a significant impact on policy formulation and decision-making. Instead, they have cocooned themselves with a commitment to a technocratic role in transport planning and have restricted themselves to the relative comfort of expert advice. By adopting such a low risk strategy their impact has been lessened, but their survival may be assured. This contrasts with the more political view taken in planning where the attractiveness of the politics bandwagon has proved irresistible, and planning analysis has become explicitly political. Ambrose (1986) describes this 'slide' of British planning into the political arena with the role of impartial technical advice being marginalized. The choice is not easy. The limitations of transport and planning procedures are manifest and the decision may be merely whether marginalization of transport planning results from positive action through becoming overtly political, or from inaction through maintaining the current *status quo*.

## The Limitations of the Transport Planning Model

The basic structure of the transport planning model (TPM) has proved robust and long lasting. The four-stage aggregate model, as originally developed in the USA in the 1950s (Chapter 2), has been the bedrock upon which analysis has taken place. Its value has been its ability to examine the city and region at the aggregate level and to establish relationships between a given land-use pattern and travel. The existing situation could be modelled through the four linked submodels (Trip Generation, Trip Distribution, Modal Split, and Trip

Assignment), and current traffic problems could then be identified. The modelling process would then be used to predict overall travel demand for the forecast year, and alternative transport strategies would be developed. It was probably the classic example of the systems approach to analysis, and represented the positivistic, descriptive and prescriptive dimensions which epitomized such an approach. The rigidity of the approach has been both a strength and a weakness. Its strength has been in the logic of the process and the representation of the ways in which decisions are made, and it also provided a framework within which transport options could be tested. However that framework has also been a source of criticism in that it has acted as a straitjacket with increasing concerns over the relevance of such an approach to analysis. The TPM is still extensively used in part or in its entirety, not because it is ideal, but because practitioners are comfortable with it and because no adequate replacement has actually been proposed. It is a tribute both to the TPM's robustness and to the inertia within transport analysis that such a situation exists.

The limitations of the conventional TPM are well known (Supernak, 1983; Polak, 1987; Atkins, 1987; Supernak and Stevens, 1987), and some of the main points are highlighted here.

### THE THEORETICAL BASIS IS WEAK

The TPM is empirically based and designed to predict travel on the basis of establishing relationships that link travel with socio-economic and other variables. This positivistic approach is data driven and makes no attempt at understanding the behaviour of people. As such it is the definitional and measurement issues which become important rather than understanding the reasons why people have to travel. It also explains why much research effort has been directed at improving the methods of analysis rather than the underlying theory of travel. The concept that social behaviour can be explained by physical laws is an attractive one and one that transport engineers are comfortable with. Practical concerns over whether changes can be predicted on the basis of empirical relationships take precedence over the more scientific ideals of understanding travel. Transport is one issue which cuts across theory and practice, with the pragmatists being in the ascendency over the theorists. Transport analysis has been seen as supporting decisions which have to be taken on investment and subsidy which often involve large sums of public expenditure. Operational approaches have been dominant over the scientific models which may increase our understanding of urban phenomena. Social scientists in the 1980s have tried to redress the balance with considerable research being directed at new approaches to transport analysis (see pages 136–154) but no alternative general theory seems to have been developed and the operational, pragmatic, predictive models are still in almost universal use.

The established structure of the TPM has maintained its position for the lack of an acceptable alternative and for the problems that social scientists face when developing alternatives in a volatile social context. Urban and transport modelling are driven by public policy which is inherently unstable and the requirements of decision-makers reflect this instability (Batty, 1989). It may be mistaken to seek an alternative paradigm to replace the existing TPM, even if there was an obvious candidate. More appropriate may be the desire to realize explicitly the limitations of the conventional TPM, and to seek to make improvements whilst at the same time exploring other means by which advice can be given to decision-makers.

A highly structured approach to planning and transport had evolved in the 1960s and 1970s which was based on the rational decision model. This process of optimization allowed goals to be set, problems defined and solutions generated by searching across a sample of alternatives characterizing the solution space, with the best plans chosen, then implemented after solutions had been rigorously evaluated against the prior set of goals (Batty, 1989).

THE STRUCTURE OF THE TPM

Serious criticism has been raised against the structure of the TPM. The sequential structure with the output of the four component parts was seen as having some logic in that the decision to travel could be divided into a series of discrete stages: whether to travel (Trip Generation), where to travel (Trip Distribution), which mode to use (Modal Split), and what route to follow (Trip Assignment). However, decision-making is sometimes simultaneous in that the decision to make a trip includes the subsequent decisions. This argument may be particularly relevant where travel patterns become routinized and habits are formed. The concept that an individual is a rational decision-maker who has full knowledge of all alternatives available (including destinations, modes and routes) may trivialize the actual behavioural decisions being made, including constraints from other household members, previous experiences, linking activities and inspiration. The certainty of human behaviour structured into the four stages and the assumptions on which decision-making are based in no way reflect either patterns of behaviour or uncertainty in behaviour. The concerns in the aggregate studies are over average behaviour, but change mainly takes place at the margin and this variability is ignored.

Within the TPM, the main concern is over testing alternative transport strategies. The interactions between land use and transport are given superficial discussion, but changes in land uses, employment and population are mainly input as exogenous variables. The continual process of change within urban areas is therefore ignored and solutions to congestion are examined solely in terms of transport. The means by which the location of activities can be used to influence levels of transport demand are excluded, and the impact that rises in property prices and commercial rents might have on travel patterns has never been tackled. Similarly, the evaluation of transport alternatives is seen in terms of user benefits, principally the savings in travel time. The more sensitive readjustment of participation in different sets of activities and in activity relocation is not covered, nor is the overall assessment of the implications of the proposal on particular groups in society. The aggregate user benefit means that more and faster travel are the principal aims of investment, and that the quality of travel and accessibility concerns are reduced to a second order of importance.

The structure of the TPM also makes it difficult to include unconventional or radical policy alternatives. It is essentially reactive and constructed so as to produce quantifiable relationships which are assumed to be the key determinants of future demand. Once these relationships have been established, between say socio-economic status and car use, they are assumed to remain stable over time. It then becomes difficult to examine, say, the impact of technology on car use or area-wide policies such as traffic calming. The structure of the TPM is always promoting more travel and greater levels of mobility. The modelling process is concerned with estimating the scale of the increase and where it will take place. Policies that explore the possibility of reducing travel demand and shortening trip lengths seem to be an anathema.

Data have always proved to be a strength and a weakness in the positivist approach, as on the one hand the models are data driven, but on the other hand there are never enough data. Most TPMs are static and are calibrated on one set of cross sectional data. Again, assumptions have to be made on whether these data are representative and whether one can predict behaviour from such information. Before and after data and various forms of longitudinal studies allow some of these assumptions to be tested (see pages 150–151). But the weakness of the primary data inputs to the TPM is not often acknowledged.

## FORECASTING

Forecasting has proved to be an 'Achilles' heel' of the TPM and it seems that increasing technical sophistication has not resulted in corresponding increases in their accuracy. Part of the explanation may relate to the theoretical and the structural issues, in particular the aggregate scale, the sequential structure, and the data driven nature of the models, but more important than these is the assumption that the future can be predicted with any degree of accuracy. Two crucial issues in all TPMs have been the assumption of stability in model coefficients over time and the assumption that variables excluded from the model will not be instrumental in modifying travel behaviour over time.

Not all significant variables can be specified in the model; there are strong assumptions made on the quantification of variables (principally on the value of time and life, and more recently the environment), and even the primary unit of study (the trip) is not a clearly defined object. Many types of movement are omitted from the TPM, such as short walk trips, interchange trips and other types linking movements, as the trip is normally defined as the full journey from one origin to one destination rather than the component parts of that trip. For example, the walk to the bus stop, the ride on the bus, the transfer to rail and the walk to the destination are all considered to form one journey. But how many trips, what mode is used, and what are the constraints on that journey? If the person stops to buy a paper or a sandwich on the journey, what happens to all the definitions agreed above? Added to this complexity is human fallibility as recall on trips made is often not accurate.

The use of activities as a framework within which to place trip making has resulted in better levels of recall (some 15 per cent increase in activities) as all time is accounted for (Jones *et al.*, 1983). Similarly, travel is averaged over the week to find a stable pattern with most effort concentrating on the journey to work. Patterns of travel do vary between days of the week (Banister, 1977), but with much greater flexibility in the use of time that variability has increased with the complexity of lifestyles. Similarly, to focus attention on the work trip oversimplifies travel patterns in the 1990s as work now accounts for less than a third of all trips (Department of Transport, 1988).

The conclusions reached by Evans and Mackinder (1980) on the predictive accuracy of British transport studies are revealing (see pages 30–31). The average forecast errors over a ten-year period are 12 per cent for population and 13 per cent for employment, whilst car ownership and household income levels were overestimated by 20 per cent. Highway and public transport trips were overestimated by 41 and 32 per cent, indicating that in addition to forecasting errors in the exogenous variables, there were specification errors in TPMs. In general, 'for none of the forecast items would an assumption of no change have yielded markedly larger forecast errors'. Similar conclusions have been reported on with respect to US experience (Institute of Transportation Engineers, 1980) and for Europe (European Conference of Ministers of Transport, 1982).

It is not just the inaccuracies in the expected growth in exogenous variables (for example population and employment), but in the structure of the four-stage modelling process itself. Crucial to the positivistic approach is the notion of the best fit model. The model is calibrated for the existing situation and once these data have been adequately represented, the strong assumption is made that coefficients remain constant over time. Errors can infiltrate the process through mis-specification of the base year model, through measurement inaccuracies, and through forecast errors in the exogenous variables. It is essential that in all modelling some form of validation is carried out, either through before and after surveys, or through the checking of model performance against an independent data set.

This point has been returned to in the National Audit Office's investigation into road planning (1988), where again it was acknowledged that traffic forecasting is not an exact science. Possible sources of error included human and measurement errors, statistical errors, modelling errors, prediction errors on specific routes, external factors (such as GDP and changes in fuel prices) and social factors (such as car ownership growth). Taking these factors into account, the Department of Transport considers that reasonable agreement between forecast and actual traffic flows 'can be assumed if their forecast is within +/– 20 per cent of the actual flow on the road one year after opening' (National Audit Office, 1988, para 4.1).

The explanations for the errors are not hard to find in the TPM, and the standard procedures used by the Department of Transport only consider trip reassignment when this is just one of a series of readjustments resulting from the construction of a new road. Other changes include travelling at a different time, the use of a different destination, the greater use of the car, and the generation of new trips. Conventional procedures use a fixed trip matrix and growth factors on flows. The implications are clear, namely that on all fronts there are likely to be differences between forecast and actual trip levels because of misspecification of the problem.

Even the tests used to check accuracy may be misplaced as the similarity of predicted and actual flows may have occurred by chance rather than the accuracy of the modelling process. The National Audit Office examined 137 sections of road opened since 1980 and found that in about one-half of the case predictions were under or over estimated by 20 per cent. The question is does it matter? The answer here must be in the affirmative as it influences the decision to build the road, the standard to which that road is built, and the opportunity cost of the capital.

The Public Accounts Committee (PAC, 1989) was even more critical and listed seven points which should be addressed by the Department of Transport (table 6.2). They also commented that the method used by the Department gives a misleadingly favourable view of the under or over estimation of traffic. In short, there seems to be considerable concern over the TPM and the forecasting process on all fronts. In addition to these theoretical and technical issues, there has been an unprecedented attack on planning as a legitimate process, with the switch away from long-term strategic analysis to shorter-term market led strategies. This is a move away from the comprehensive city level analysis with the intention of understanding how cities work and the relationships between transport and land use towards much more specific and partial analyses. These changes have been paralleled by the reorganization of local government and the abolition of strategic planning in the metropolitan counties (and London), regulatory reform, privatization, and the primary concern over efficiency in transport and reductions in levels of public expenditure (Chapter 4).

TABLE 6.2. The recommendations of the Public Accounts Committee.

1. Evaluation is needed of the effectiveness of completed road schemes together with monitoring the outturn of the original forecasts and appraisal assumptions.

2. Evaluation of the environment is important in reaching decisions on the roads programme.

3. Traffic forecasts must be improved and the Department of Transport's reluctance to accept that there is a serious problem 'verges on complacency'.

4. The M25 is a particular problem and the inaccuracies of the forecasts cannot be fully explained by economic growth. There are also generated and redistributed traffic, and regional factors.

5. Support any moves to speed up the road planning process.

6. The Department of Transport should assess the additional costs to industry, commerce and the Exchequer of not providing extra capacity.

7. The Department of Transport should take account of the wider and more strategic consequences of building new roads.

## COMMENT

Despite these strong criticisms, 'improvements' have been made to existing approaches rather than shifting to some new paradigm. Such a paradigm shift does not occur through overt criticism of existing practice, as this tends to result in professional barriers being erected and a defensive protection of established practice. To achieve a paradigm shift, the critique must be made in terms which the existing practitioners can accept and relate to. This means that it must be made in terms that are familiar to them.

In transport, such an opportunity has now occurred. The TPM was primarily developed to allocate growth in population, in economic activity, in income, and in car ownership. It was concerned with the provision of more transport infrastructure through the expansion of the road network and, more recently, improvements in public transport supply. As Hutchinson succinctly states (1981)

- The provision of large amounts of road capacity encourages the use of cars for commuting, but more importantly encourages long trip lengths between home, work and other activities.

- Public transport demands are influenced by car ownership, but the principal determinants of public transport use are employment concentration on public transport routes combined with high parking charges or parking space constraints on these employment locations.

- The principal dimensions of transport demand, the efficiencies with which these demands can be satisfied, are dictated by factors external to the transport sector such as the housing market, household structure and income, and employment location decisions.

- There is a great uncertainty in the social, economic and technical environment of transport systems, and transport plans must be sufficiently robust and adaptable to changes in the transport environment.

These perceptive comments made over ten years ago are still relevant today and reflect the inbuilt resistance to radical change. It has only been with the impact of renewed growth in traffic in the 1980s that some convergence has taken place on the view that increasing supply will not provide a solution. Travel demand overall is increasing at a rate which is substantially faster than effective capacity will (or can) be expanded (Goodwin *et al.*, 1991). At about the same time the Department of Transport (1989*a*) also published its *National Road Traffic Forecasts* which confirmed the view that the expected growth in traffic over the next 30 years would be in line with growth in Gross Domestic Product. Traffic would grow by between 83 and 143 per cent. Since then there has been a near universal call for more investment in public transport and restrictions on the use of the car through physical restraint and pricing for road space.

The irony here is that similar views were expressed in the 1960s when a similar growth in traffic was experienced. These demands were dissipated in the 1970s as economic factors and global recession resulted in lower levels of growth, and other policy concerns absorbed the attention of politicians and the public. Perhaps the same situation will occur in the 1990s as severe recession replaces the buoyant economic growth of the 1980s.

## The Response in the 1970s and Early 1980s

Research has proceeded along several different dimensions in the attempt to respond to at least some of the shortcomings of the traditional four-stage aggregate approach. In the 1970s and early 1980s, there were two main strands of thinking, one concerned with improving the existing approach and the other concerned with the individual choice process.

### INTEGRATED LAND-USE TRANSPORT MODELS (ILUTM)

One of the basic limitations of the conventional TPM was treatment of land use as an exogenously determined factor. The model was primarily concerned with changes in transport demand and in the evaluation of transport alternatives. The ILUTM still took a systems approach to analysis but attempted to model the interactions between land use and transport explicitly within the modelling process, principally through various concepts of accessibility. Different approaches have been adopted. Regression models have been widely used to develop an understanding of the spatial distribution of land uses by making the dependent variable the population, employment or housing in each zone, and using measures of accessibility to determine the appropriate distribution. Different land-use variables can be modelled through sets of simultaneous equations, and the output would be a series of measures relating to the level of activity within each zone. Mathematical programming models allow the allocation of activities to each zone to be optimized. The Australian model TOPAZ (Technique for the Optimal Placement of Activities in Zones) optimizes the allocation of activities to zones by minimizing the total costs of establishment and travel, subject to the constraints that all activities have to be located and all zones are filled (Brotchie *et al.*, 1980). The main determinant of the allocation is accessibility which is included in the form of a gravity model. Similar approaches have been used in the SALOC model developed in Sweden (Lundqvist, 1984). By definition, optimization produces a solution to the specified problem, but makes no attempt to suggest the means by which that solution can be achieved, and no indication is given of the competition process between different land uses, nor to the important role that the pricing mechanism has in determining land use.

By far the greatest effort and interest has been in the use of spatial interaction models, which have strong links with the conventional TPM and the Lowry Model (Lowry, 1964). The model is comprehensive, as it covers housing, residential location, employment, shopping, journeys to work and shops, and land allocation processes. Research in the USA includes the Projective Land Use Model (PLUM) and the Integrated Transport and Land Use Package (ILTUP) originally developed by Goldner (1971), and modified by Putman (1983 and 1986).

In Britain, apart from the pioneering research by urban modellers in the 1970s (Chapter 3), most recent attention has come from Echenique and Mackett. MEPLAN (Echenique *et al.*, 1990) is a representation of the spatial economy of a region and it explicitly links the interactions between land uses and transport. In each zone, MEPLAN estimates the equilibrium price of land or building resulting from the demand created by activities in the zone and the supply of land or building at a particular point in time. The equilibrium price of transport is also calculated in the form of accessibility, and this forms the main input to the model for the following time period. Land uses determine the demand for transport and accessibility determines the price and hence the demand for land or buildings.

The LILT model (Leeds Integrated Land Use Transport Model) modifies the basic Lowry structure to include modal split, capacity constraint and car ownership (Mackett, 1985). More important though is the introduction of a semi-dynamic structure to the models with a progression from the base year over a period of time, typically in five-year periods, with lag mechanisms to link land use and transport at various points in time to ensure that all constraints are met.

The culmination of the research on the ILUTM has been the ISGLUTI (International Study Group on Land Use Transport Interaction), set up by the Transport and Road Research Laboratory in 1981. This comparative study takes seven predictive models and two optimizing models, and carries out a series of comparisons on them (table 6.3). It was felt that many of the models had similarities and a common theoretical base, with the extensive use of entropy maximization procedures, spatial interaction models (gravity models), and the location of basic employment as the main driving force (Webster and Paulley, 1990). A range of common land-use and transport policies have been tested on the different models, and some of the models have been calibrated with different data sets to determine the differences in output. The difficulties of international collaboration and the problems of transferability of similar models have created problems, both for the interpretation of policies and for the evaluation of the output (Mackett, 1990). All of the models are access driven as land-use changes respond to changes in accessibility, and transport is then matched to this new distribution. However, with high levels of car ownership and with industry becoming more footloose, the question arises as to whether access is the correct mechanism, as it is interpreted only in a physical sense (that is the generalized cost of reaching the desired location). Other factors such as the availability of land, land prices and the development potential of land are all important in the location and travel decisions, but these are input exogenously.

Some of the limitations of the TPM have been met by the ILUTM. But the complexity of the land development processes, travel decisions and the rapidly changing forms of industry, of population structure, of lifestyles, and of the use of time all contrive to make progress difficult, if not impossible.

*Simulation*

Simulation provides another novel way to model the decision process, often at the level of the individual or household. Simulation specifies a set of possible events and the likely outcomes of each event for one or more variables, and these events are then simulated exogenously with the repercussions of each being assessed. Monte Carlo methods are often used to drive the process with the random numbers generated being compared with the cumulative probability distribution, which determines the outcome of the process. Aggregation takes place so that the macrolevel impacts can be assessed.

The MASTER (Micro-Analytical Simulation of Transport, Employment and Residence) model is one such application (Mackett, 1990). A given set of supply data is input to the model and the various alternatives in the sequences that a household can follow are specified. The model then proceeds to determine the impact of demographic processes, of migration, residential location choice, economic activity and lifecycle, change of job, and finally transport.

TABLE 6.3. Participants in the ISGLUTI Study.

| Country | Organization | Main Members | Name of Model and Year First Developed |
|---------|-------------|--------------|------------------------------------------|
| Australia | Commonwealth Scientific and Industrial Research Organisation | Brotchie Sharpe Roy | TOPAZ 1970 |
| FR Germany | University of Dortmund | Wegener | DORTMUND 1977 |
| Japan | Universities of Tokyo and Nagoya | Nakamura Hayashi Miyamoto | CALUTAS 1978 |
| | University of Kyoto | Amano Toda Abe | OSAKA 1981 |
| Greece | University of Thessaloniki | Giannopoulos Pitsiava | |
| Netherlands | University of Utrecht | Floor | AMERSFOORT 1976 |
| Sweden | Royal Institute of Technology | Lundqvist | SALOC 1973 |
| UK | University of Leeds | Mackett Lodwick | LILT 1984 |
| | Marcial Echenique and Partners | Echenique Flowerdew Simmonds | MEP 1968 MEPLAN 1985 |
| | Transport and Road Research Laboratory | Webster Bly Paulley | |
| USA | University of Pennsylvania | Putman | ITLUP 1971 |

*Source*: Based on Webster and Paulley (1990).

Not every household or household member passes through all stages in the microsimulation at each time period, and a wide range of related transport decisions can be analysed, such as the purchase of the car, the use of that car and the modal choice decision.

COMMENT

These two developments, ILUTM and simulation, have advanced the TPM in several important respects. The links between land use and transport have been made explicit, with the realization that relocation of activities will not necessarily lead to more travel but could result in household relocation, moving job, and the substitution of one activity with another. However, by concentrating on the physical factors, the economic and social dimensions may be demoted in their importance. Microsimulation can capture both the economic and social factors by exploring housing costs, the costs of job relocation and the car acquisition process. Secondly, these models have moved away from the comprehensive claim of the conventional TPM towards examining the interactions between transport and other factors in the decision process. This marks a greater realism in thinking, and it also provides a closer link with the requirements of policy-makers.

The mid-1980s probably announced, at least unofficially, the final demise of the conventional TPM. That is not to say that it is not still being used, but the concept of an all embracing approach to transport planning has disappeared to be replaced by a more focused approach which takes particular policies or problems for analysis. The strategic approach, involving extensive data collection and modelling, is not seen now as an essential prerequisite to major decisions in the transport sector. Analysis is likely to be more selective using simpler sketch planning approaches to pick out trends. In addition, statistical and limited modelling analysis will be based on a clear statement of policy objectives. The rational comprehensive approach with the systematic consideration and evaluation of transport alternatives has been replaced by short-term broad based analyses carried out at the strategic level and supplemented by detailed local studies. It is here that broader based modelling exercises and simulation studies, such as those outlined in this section, have an important role to play.

DISAGGREGATE BEHAVIOURAL MODELS

To complement the large-scale integrated land-use transport models and the simulation studies, extensive research has also been carried out at the micro-level on the individual choice process. The basic purpose of these behavioural models is to predict the particular aspects of travellers' behaviour most relevant to the planning of transport facilities (Daly, 1981). This approach has arisen out of the realization that aggregate travel phenomena are the result of disaggregate decision making by a large number of individuals. If the decisions made by each individual can be modelled and understood, then the approach can be called behavioural with a corresponding theoretical foundation. Three basic approaches have been developed, one based on utility theory and the notions of rational choice behaviour, the second uses the principles of psychological choice theory, and the third places travel decisions in the framework of activities. This section briefly outlines and comments on each of these approaches.

## Disaggregate Utility Models

Disaggregate Utility Models evolved from micro-economics and the theory of consumer choice developed in the late 1960s (Lancaster, 1966). Individuals (consumers) act so as to maximize some benefit or utility, with the individual being represented as exercising his or her choice over the full range of available options, limited only by the constraints of time and money. In transport, this choice was normally represented as discrete alternatives between mutually exclusive modes, such as the choice of the bus or the car for the work journey.

This approach epitomizes the rational choice model where individuals have full information on all the alternatives and where they can rank those alternatives according to their own preferences so that the 'right' choice is made. As such it suggests how trips should be made rather than how they are actually made. In its operational form certain simplifications are made, with the random utility model assuming that individual trip makers weight the various aspects of travel differently, where these weights are assumed to be randomly distributed according to particular probability distributions (Hutchinson, 1981). Different probability functions lead to different choice functions, with the most commonly used formulation being the logic choice function where the weights are assumed to be Weibull-distributed. Utility functions are developed for each mode and disutility coefficients estimated so as to maximize the ability of the model to explain the modal choice decisions of individuals (Domencich and McFadden, 1975). Although modal choice has been the main focus of attention, similar approaches have been used to model destination choice, car purchase decisions and car pooling.

## Attitudinal Models

Attitudinal Models shift the focus away from the concepts of utility maximization to those of satisficing behaviour. The individual is making choices in a situation of partial knowledge and when certain thresholds are reached, action will take place. Attitudinal methods are more subjective in their construction and relate to behavioural intentions and consumer choice. Apart from having a modelling role, attitudinal methods have also been used to improve the understanding of individual choice through exploring the links between attitude, preferences and behaviour, and they have provided the means by which attitudinal variables can be included in existing models. Values can be placed on those variables (other than time and cost) considered important in individual choice. These could include measures of comfort and convenience, as well as service reliability and the perceived safety and security of different modes of transport.

Attitudinal methods mark an important change in thinking as it is implicitly accepted that revealed demand may not be the only measure of demand. Almost all analysis takes the trips made as the dependent variable for which explanation is sought. Attitudinal approaches assume that behaviour and travel are wider concepts and an exploration of the decision processes can help explain behavioural intention as well as actual behaviour. The underlying assumption here is that attitudes influence behaviour with an implicit link between an individual's preferences and his or her choice of mode. So if attitudes can be measured, some preference ranking can be obtained from which actual choices and behaviour can be estimated. The difficulty here is that preferences are broader than actual choices and so behavioural intentions may overestimate actual demand.

The weakness in the above argument is the assumed causality in sequence from attitude to preference to choice and behaviour. This oversimplifies the complex psychology of choice. Behaviour also influences attitude and the two elements may interact to maintain consistency between attitude and behaviour (a cognitive dissonance effect). Individuals modify their attitudes to justify what they actually do, so that they have a positive attitude to the car if they use that car. Similarly, there may be some mutual dependence between attitudes and behaviour. Learning theorists argue that through a process of stimulation and response there is reinforcement (positive or negative) which results in routinization of behaviour and the formation of habits (Banister, 1978). Choice is treated as a process and *not* an event with behaviour being modified through adaptation and learning. The overriding concern of most methods over accuracy of output and predictive capability is replaced by a more modest objective of testing different behavioural assumptions and understanding behaviour. The concern is more with diagnostics than prediction.

However, while attitudinal methods do provide a range of approaches to measure the strength of beliefs and the likely behavioural response to change, their real contribution will only be apparent after extensive applications, and this has now been achieved in transport, primarily through the use of stated preference methods (see pages 149–150). Whether compensatory or non-compensatory structures are used, testing and validation are the only means by which the strengths of attitudinal methods can be assessed.

## Activity Based Approaches

Activity Based Approaches present an important theoretical advance as they view travel as a derived demand rather than a direct demand. This means that people only travel because of the benefits that they perceive at their destination, and that little travel is undertaken for its own purpose. Following on from that premise, the activity based approaches assume that travel is one of a set of activities which people participate in and it provides the link between activities which do not take place at a particular location.

Much of the pioneering work was carried out in Sweden (Hagerstrand, 1970) where a general framework for human interaction has been developed through which travel behaviour can be analysed. The aims of this framework are to demonstrate the activity opportunities and constraints which face particular individuals or groups of people. Activities are constrained by their location (spatially), by their opening hours (temporally) and by whether they can be visited as part of a 'tour' (complementarity). People are constrained by their physical and economic conditions (personal constraints), by their work and school hours (time constraints), and by linkages with members of their own family and others (coupling constraints). In addition to these two sets of limitations there are also authority constraints which are imposed from outside (for example bus timetables). Travel is assumed to be limited by these constraints and so the space time prisms can be constructed within which any individual's opportunity set lies. It is here that transport plays a crucial role as the shape and size of this prism is often controlled by the availability of a car. In essence, such a constraints-based approach is a physical representation of society which explicitly portrays travel and non-travel activities. Within this framework the dynamics of behaviour can be investigated, and it allows an understanding of the types of specific adjustments which might take place as a result of a policy change (Carlstein *et al.*, 1978).

Activity based approaches have been extensively used in a variety of specific applications

- Diary studies have explored the use of time and space over time, in particular locations, by different social groups, and subject to particular constraints (for example car availability).

- Time geographic accessibility studies have combined all the concepts into a modelling approach based on two-dimensional space and time. A simulation model (PESASP) provides the answer to the question as to whether an individual's activity path is physically compatible with the constraints imposed, and it shows the number of possible combinations and the different sequences by which activity programmes can be achieved (Lenntorp, 1981).

- Interactive gaming approaches simulate travel and other aspects of behaviour over particular periods of time. These approaches can either examine the responses of families to particular changes in policy or focus on selected issues such as the environment and the household budget allocation decision. In each case the respondents, usually members of the household, present their current set of activities, prior to a change being introduced followed by a discussion and the resolution of the new set of activities (Dix, 1981; Jones *et al.*, 1983).

- Activity scheduling models include an explicit dynamic element and examine the linking of sequences of activities through trip chaining or the ways in which household schedules are formulated. In all of the scheduling approaches, time budgets and the use of time are considered to be the driving force in terms of the sequences and the range of activities that households can participate in.

COMMENT

The disaggregate behavioural models are much more modest in scope than the conventional TPM and they are based on different views of how individuals make decisions. In this sense, they are behavioural, but a general theory of travel behaviour has not been developed. Each approach sets out a series of assumptions which allow an analysis to take place, but reality is a mixture of the following three facets. Individuals are not utility maximizers, nor preference satisfiers, nor completely limited in their activities by the constraints of time and space, but a combination of all three.

Where these approaches make a significant contribution is in the extended range of policy options which can be examined and in the move away from the obsession with long-term prediction towards a short-term view of reactions to particular policy changes. Their concern is over the impacts on particular social groups and it is accepted that travel decisions are complex. However, these methods have not gained widespread acceptance in practice. Heggie (1978) argued that the responses suggested were at odds with empirical evidence on behavioural responses to changes in fares and levels of service, and that there is still an underlying ethos which wanted a statistical explanation rather than a behavioural one.

Despite these limitations, several important policy issues can be addressed and have been analysed by this first generation of behavioural models (1970–1985). These are summarized in Hutchinson (1981)

- changes in family structures and lifestyles and the related impacts on transport and housing markets;

- the ageing of the population and the impacts on transport needs as they relate to other markets such as housing, health services and recreational opportunities;

- the changing employment bases of urban areas and the implications for fixed route, centrally-focused transport facilities;

- increasing energy costs and the impacts on residential, job and retail markets and on transport demands and needs;

- the reductions in economic activities and the ability of governments to continue to finance the deficits of public transport systems;

- the impacts of the expected innovations in computer communications technology on household and institutional activity patterns and organizations.

Available modelling techniques are not capable of completely responding to these issues and observed behaviour may not be the most appropriate way to investigate likely future response. As the uncertainty, complexity and flexibility of individual actions has greatly increased, new methods must be developed to investigate these impacts.

## The Response Since 1985

In the 1960s when urban planning and transport analysis were in their infancy, a series of landmark conferences was held, principally in the USA (Chapters 2 and 3). In the mid-1980s a second series of important conferences was also held to mark the switch in transport planning away from the primary concern over the rational comprehensive approach towards the use of a much wider range of analysis tools. Conferences were held in the UK, Japan, Australia, France, Canada, as well as the USA. The output from these conferences was a restatement of the current position with respect to travel demand forecasting and transport planning analysis. However, the more important output has been the attempt to sketch out the new agenda and requirements for analysis.

Goodwin (1991) presents a clear summary of the usefulness of today's approaches:

- Assignment models to predict traffic flows for trunk and local schemes; they assume that a scheme will only influence route choice within a fixed total number of trips, i.e. that travel demand is only marginally responsive to transport supply or costs, and are calibrated in order to reproduce an approximately known level of flows in a base period, using traffic counts.

- Four-stage models for big complicated schemes in large urban areas; they allow destination, mode and route choices to be influenced by transport supply, but usually not the frequency or structure of trip making, (except for committed land use developments generating a pro-rata increase in trips) or in time of day of travel. They are calibrated in order to reproduce approximately measured patterns of travel on a notional single ('average') day in base year, usually using household surveys and counts.

- Aggregate procedures using implicit or explicit elasticities or other behavioural parameters to forecast national traffic levels and aggregate public transport usage. They are typically based either on a non-lagged regression of a statistical time series over several years, or discrete choice modelling more or less similar to logit models of mode choice, or sometimes ad hoc interpretations of miscellaneous data to hand.

• In recent years, surveys often use stated preference techniques to forecast usage of new facilities or service modifications.

• A subtle process of judgement designed to adjust, constrain or 'reinterpret' model forecasts especially when in the initial stages of a forecasting exercise there are odd or counter-intuitive results. This judgement is a necessary part of any forecasting process; it should be remembered however that the fundamental credibility of a model, and the insidious influence of previous modelling experiences in actually forming the intuitive basis on which the judgement is built, still need to be consciously examined.

The struggle faced by many researchers was that even though they had considerable reservations about the established procedures and there was no lack of new ideas (see pages 136–145), little change took place. It seemed that research which did not easily fit into the established order was marginalized. Thus the opportunity for improvement or for diversification was rejected in favour of the traditional approaches to analysis. Goodwin (1991) has strongly argued that much more research has been done than is appreciated, and that although it was of direct relevance to the policy needs of the time, it was ignored because it did not fit into the established framework. He goes on to suggest that much of this 'marginal' research should be reclaimed.

As stated earlier (see pages 125–129), transport planners have been resilient in avoiding many of the theoretical arguments presented by political economists and others in the 1970s, and their reaction was to seek refuge in more technical arguments with which they were familiar. Consequently, the basic framework for transport analysis has remained unchanged for 25 years, and it has only been recently that greater flexibility has been apparent. Two strands of research have continued through the 1970s and 1980s, and these have been outlined in the previous section. Five other more recent developments have taken place, and together they could form part of the possible renaissance of transport planning as they each reflect new techniques and ideas which have made the transition from research to practice.

INTEGRATED TRANSPORT STUDIES (ITS)

Integrated Transport Studies (ITS) have been carried out in about 20 cities in Britain (1987–1991) as a base for guiding transport strategies over the next 20 years. Their origins derived from the new requirements of the Department of the Environment (1989) for clear statements on transport policy in the Unitary Development Plans, particularly when a case was being presented for infra-structure investment in new light rail systems.

The procedures developed (May, 1991) require a statement of vision for the town or city which covers economic, environmental, and quality of life aspects, together with the specific transport objectives to meet the vision and a realistic statement on the likely finance which is available. Against these objectives, potential transport problems can be identified and this requires a prediction of future conditions, but acknowledgement of the inherent uncertainty that exists. Strategies which cover both land-use and transport options are specified and evaluated against the full range of objectives. The best elements are then combined in a preferred strategy.

The basic differences between the ITS and the TPM are that the time horizon is reduced to the medium term, objectives are very clearly stated at the beginning of the study, and a wide range of strategies are tested. The whole study takes between three and six months, and there is no requirement for detailed model building, data collection and calibration. Strategic sketch planning models are not new but they have suddenly come back into fashion. The structure of the Birmingham Study (figure 6.1) has been used as a role model for other studies (Jones, *D. et al.*, 1990). The preferred strategy is fully costed and evaluated through multi-criteria analysis, which relies on both model output in terms of meeting expected demand and professional judgement.

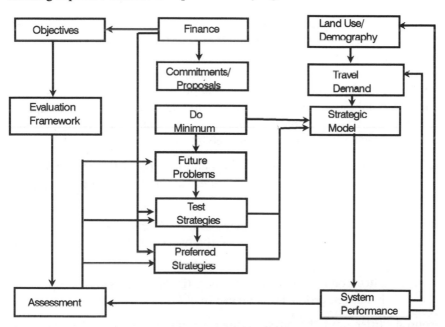

FIGURE 6.1 The Structure of the Birmingham Study. (*Source*: Jones *et al.* (1990))

The ITS have rapidly obtained respectability and their use is likely to increase as the approach is familiar to both transport planners and politicians. The basic question is one of balance, whether the ITS is too superficial or whether it does encompass the main alternatives and investigate them to a sufficient depth so that full consideration is given to uncertainty. Only time will provide the answer to this question.

EVALUATION METHODS

Evaluation methods provide a second area where a radical rethink is taking place, partly as a result of concern over existing economic based approaches and partly as a result of the growing importance of externalities, principally the environment.

Comparability is being sought between evaluation for different modes of transport, the value of time (which accounts for about 85 per cent of user benefits in road evaluation) reviewed, and the question of comprehensiveness in evaluation is now being discussed. The connection between strategic environmental issues and strategic transport planning is poorly developed both analytically and in terms of its presentation to decision-makers (Wood, 1992). A crucial issue here is the appropriate time for the appraisal so that it fits in with the decision process and when the expected impacts might be apparent.

The role of cost benefit analysis in evaluation is being qualified, as other multicriteria analysis methods (such as the Goals Achievements Matrix) may provide a framework within which the economic, political and environmental factors can be placed. Local authorities have made considerable use of Goals Achievements Matrix approaches to evaluation (for example London Strategic Policy Unit, 1986), but the most recent comprehensive use of a multicriteria framework was in the common approach adopted in the London Assessment Studies (Coombe *et al.*, 1990). The framework identified four main groups of impacts (and two supplementary groups), and presents a large table with one column for each option and rows for each impact.

Group 1   Economic effects on travellers and transport operators including efficiency, safety and costs (travel time and vehicle operating cost changes).

Group 2   Economic costs of implementation, enforcement, maintenance and operation (includes costs of roadworks, landscaping, land and property).

Group 3   Effects on the human environment (includes users and occupiers of buildings, pedestrians and cyclists).

Group 4   Effects on the physical environment (through brief descriptive statements on heritage and urban structure).

Group 5   Effects of the options on the Boroughs' local objectives.

Group 6   Special effects such as impacts on ecology or agriculture.

The options being evaluated in each of the four corridor studies in London – West London, East London, South London and the South Circular – were all placed in this evaluation framework and assessment was based on the achievement of objectives, the effects on problems, and value for money. The options involving public transport were as far as possible assessed on the same basis as the road options.

## STATED PREFERENCE AND CONTINGENCY VALUATION METHODS

Stated Preference and Contingency Valuation Methods are two new techniques which have been incorporated into standard practice. Using interview techniques and attitudinal questionnaires it is possible to elicit preferences for alternative choices. The process can either be carried out in a non-compensatory structure where the alternatives are eliminated after an attribute by attribute comparison (for example Tversky's Elimination by Aspects Model) or in a compensatory structure which requires a trading off of higher levels of satisfaction on one attribute against lower levels on others, unless the cost in increased (Kroes and Sheldon, 1988; Louvière, 1988). These methods have now been widely used to assess preferences for different modes, and to place values on quality of service variables. They are particularly useful where new or radically different proposals are being assessed, and where past trends are unlikely to give a real estimate of future demand.

The contingent valuation method (CVM) goes one step further and actually asks respondents to state a price that they would be willing to pay for a particular alternative, or what they would be willing to receive by way of compensation to tolerate a cost. Again, CVM methods may be the only choice of method in particular situations, such as valuing the environment. The problems are similar to stated preference in that there is not a direct correspondence between the hypothetical market and the real market, and there is no empirical symmetry between willingness to pay (WTP) and willingness to accept (WTA). Conventional economic theory suggests that an individual should show indifference between WTP and WTA. Preferences and contingent valuation both tend to overestimate demand, and this limitation means that validation is important. Nevertheless, CVM is one of the only methods available for exploring the different means to assign monetary values to the environment.

> Many costs and benefits are measured directly in money terms: for example, savings in expenditure on resources, and sales revenue. Where they are not (e.g. travel time saved, noise and other forms of pollution, and broad managerial or political factors) costs and benefits can sometimes still sensibly be given money

values, often by analysing people's actual behaviour and declared or revealed preferences. These imputed money values can be used in the appraisal as if they were actual cash flows. Other factors which cannot be valued should be listed, and quantified as far as practical, making it clear that they are additional factors to be taken into account. It can sometimes be helpful to calculate the value which such factors would have to take for the net present value of a scheme to turn positive or negative . . .

Account sometimes needs to be taken of the value which individuals may place on the possibility of using a service or visiting an attractive area even when they do not currently use the service or make visits. (UK Treasury, 1984, paras 33 and 34)

These statements suggest that actual behaviour and revealed preferences can be augmented by surrogate markets (for example hedonic prices) and hypothetical markets (for example contingent valuation), as well as option value. Again, the crucial transition of concepts from research to practice has taken place, and a broader interpretation of preferences and values can be included in selecting alternatives.

DYNAMIC ANALYSIS

Dynamic Analysis has become the focus of some research in the social sciences, even though earlier attempts at longitudinal data analysis were inconclusive (Wrigley, 1986). This may have been because the earlier efforts were directed towards the collection of data rather than its analysis. More recently, repeated cross section data sets have become available so that the output from static analysis can be verified and this in turn has led to dynamic modelling becoming of interest to transport researchers. Richer data sets involving diaries and panel surveys have also allowed the dynamics of travel behaviour to be explored (Jones, 1990). The practical use of these data sets and the associated analytical research has been more apparent in the Netherlands and the USA than in the UK (Goodwin and Layzell, 1985).

The theoretical base for such an approach is strong. Individuals do not make decisions in isolation, and subsequent decisions are always influenced by prior decisions. An individual's experimentation, learning and experience all accumulate over time, leading in turn to preferences, values, habitual behaviour, and decision rules. In addition to the dynamics of behaviour there is also the time dynamic. Behaviour is not replicated each day and activities are scheduled over longer periods of time. Dynamic travel analysis investigated both the behavioural and temporal aspects of travel decisions.

Most dynamic analysis in transport has focused on the modelling of specific aspects of the travel decision (such as route choice and departure time), and on establishing a consistent analytical basis. There are methodological problems raised by the type of data collected (mainly qualitative) and the interrelationships between variables. For example, in panel analysis much research has focused on the difficulties in obtaining and maintaining a consistent analytical framework over the panel waves. As Kitamura (1990) suggests, consideration must be given to

- possible increases in non-responses due to the fact that respondents are required to participate in more than one survey;

- problems of sample attrition;

- problems of locating respondents in multiple survey waves due to residential location and dissolution of households;

- possible decline in reporting accuracy due to 'panel fatigue';

- problems of 'panel conditioning' where behaviour and responses in later surveys are influenced by the fact of participating in the panel, and also by responses to previous surveys.

These problems are not unique to panel surveys, but are a price which has to be paid for better quality data. Dynamic analysis is still in its infancy. Although it may offer many theoretical and methodological advantages, panel surveys are probably the least likely of those presented here to be widely accepted and used because of the additional costs of data collection and the skills required for analysis.

## LARGE SCALE TRAVEL DEMAND SURVEYS

In all the developments itemized in this Chapter there has been one underlying theme, namely data. Information about travel decisions, about the city, about demographic factors, whether collected at one point in time or over a series of points in time, has featured at all stages. Good quality data are difficult and expensive to collect, and one feature of the 1980s has been the absence of the large-scale travel demand surveys which characterized the classic Land Use Transport Studies in the 1960s and 1970s.

Updating of databases has taken place, but there are now increasing requirements for new information for monitoring and system performance measurement. Technology now permits the collection of huge amounts of data on the use made of the transport system, both by cars in congested urban conditions and by public transport through the use made of passenger information systems, automatic ticketing and smart cards. However, all these data only report on actual travel behaviour and on the performance of the transport system. They do not explore the motivations for travel, household decision-making or suppressed demand.

The concept of travel data as an information source is an important one. Modelling in the traditional transport planning sense is no longer the main reason for collecting travel data and the data themselves are not suitable for input to the TPM. Database enquiries, the use of real time information, and interactive management systems are all more important concerns for transport operators and managers. The question then becomes whether this information can be used to investigate travel behaviour or what limited supplementary data are required to link the travel data collected so that modelling can take place. The survey process changes to one of database assembly and management together with the most appropriate means to interrogate that information. Surveys will only be undertaken to provide essential supplementary data and to update the data. All information will link in with the basic database structure (Taylor *et al.*, 1992).

Geographic information systems (GIS) provide the basic structure for many databases as their spatial organization allows information to be added or extracted very easily. Through an interactive graphics interface, it is also possible to 'see' the data. The analyst can also impose a particular structure on the data (for example household, zone or area) to suit his or her own requirements. This permits analysis at different scales and for linkages between the scales. The database has not been assembled for one purpose but for all users, and so this should allow for a richer analysis of interactions between sectors, as well as single sector investigation. The flexibility of GIS databases is illustrated in figure 6.2 where the linkages between modelling, the integrated spatial database (GIS) and individual supplementary databases are clearly shown.

COMMENT

One of the main problems with technical forecasts and trend analysis is the obsession with numbers, not with the underlying principles and ideas. Little thought has been given to visions of the city as a place in which to live and work, and how such a goal can be achieved. Planning and transport planning in particular have both become calculative. As Schon (1983, p. 39) puts it:

> attention to problem solving has been at the expense of problem setting – students have been taught how to put real-world problems into the molds of techniques which appear to render them soluble, but have not been taught how to inquire into whether they are asking the right questions in the first place.

There is a yearning after scientific rigour based on a premise of technical rationality which favours harder physical science rather than softer social science.

The attractiveness of the apparent ease with which quasi-scientific approaches can provide easy answers to difficult questions is understandable. There is a reassurance in numbers and an intolerance of uncertainty. Technocratic analysis such as that found in the conventional TPM gives that reassurance by assuming that travel demand at the aggregate level is repetitive

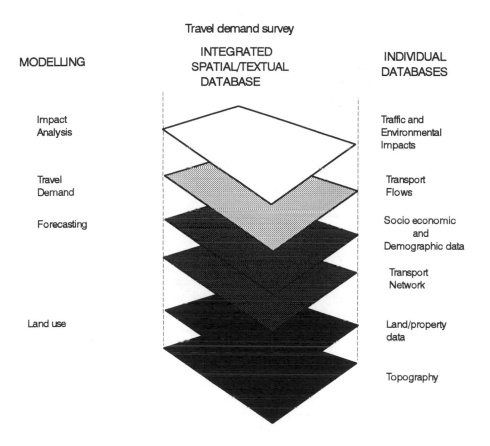

FIGURE 6.2 Transport Related Databases in a GIS Framework. (*Source*: Taylor *et al.* (1992))

and predictable.

But transport analysis does not really explore the reasons for change which relate to economic development, social policy, the globalization of economies, the increase in affluence and leisure, and the use of technology. In one sense, the 1989 National Road Traffic Forecasts were a watershed (Department of Transport, 1989*a*), with the realization that more roads would not solve the problem, but the real problem is one of single sector vision which does not draw linkages between transport and what is happening in society as a whole (table 6.4). The case is not to dispense with quantitative methods, but the myth that they are unbiased arbiters in prescribing proper planning actions. 'They must be moved away from centre stage and replaced by a broader more critically oriented approach' (Richmond, 1990, p. 52).

TABLE 6.4. New directions in transport planning.

1.  Move away from trend-based extrapolation to richer social analysis based on linking transport to what people do and how industries operate.
2.  Change the balance in evaluation so that greater weight is attached to qualitative factors and ecological arguments.
3.  Move away from 'objective' factors in analysis (for example cost and time) towards the acceptance of subjective valuation and political rationality. The primary concern over time saving should be replaced with measures of improvement in the quality of life.
4.  Greater realism in understanding people's rationality in what they do, particularly in developed high-income economies where the marginal utility of income may be limited.
5.  Abandonment of single sector analysis in favour of integrative studies which relate to the city or locality as a whole

## Renaissance in the 1990s?

Almost all analysis in transport planning has been based on modelling approaches which crudely represent the economic processes in terms of the demand for movement and the available supply of infrastructure and modes. The links between the models and the actual structure of activities in cities and regions are tentative, and a direct correspondence is assumed to exist between land use and travel. The treatment of travel is as a physical concept. These models, both transport and land use, represent a fusion of gravitational concepts underpinning spatial interaction with macro-economic theory as reflected in input-output and economic base analysis (Batty, 1989). Many of the subsequent developments outlined here have still only been used as applied research tools rather than as policy-making instruments. There has been little interest from researchers to study substantive new issues and new policy questions from a modelling perspective. As Batty (1989) comments, modelling has acquired a life of its own and has become institutionalized. It has retreated from the volatility of public policy-making into its own cocoon and it has not responded to the challenge of social change. Batty explains this essentially defensive strategy in terms of the eclectic set of disciplines from which urban and transport planners emerged, and from the small number of researchers involved in the field. Despite these limitations, the research output has been impressive, at least until the mid-1980s, although the greatest progress has been made in analysis methods rather than as an input to the public policy debates.

For more than 20 years, transport planning has been on the defensive as priorities moved from planning growth to the management of decline in urban areas, and away from providing for the car to managing the use of the car. Transport is no longer seen as a means to distribute growth in the urban area, but as a crucial lever in regenerating economic growth in areas where the traditional economic base has been ravaged in the transition to a post-industrial urban society. The critiques of urban planning and transport raised in the 1970s (for example Sayer, 1976) which attacked the positivist approach as being unidirectional in causation and not involving dynamic processes and feedback, now seem irrelevant to the agenda which has evolved. Traditionally, competition was seen in terms of substitution. For example, a new transport mode competes directly with existing modes for patronage. New forms of communications, information transfer and the restructuring of industry result in increasing linkages between alternatives and additional travel being generated by a wide range of modes. Processes have become much more complex, and the realization that the action-reaction cycle has many new dimensions forms one of the main challenges of the 1990s.

The questions raised are fundamental. It is unclear whether models in their traditional sense can accommodate the ideological culture of self-interest which characterized the 1980s, and the technological and demographic changes which are taking place in society. The representation of urban systems in spatial terms may only present part of the picture as technological and demographic change are essentially aspatial, but have spatial outcomes in terms of distribution. The culture and attitudes of people, business, industry and governments all influence what actually happens. All forms of planning have been exposed to the new competitive requirements in terms of their own structure, their accountability and their range of activities. Not only have the questions changed, but so have the answers.

Space is still a crucial determinant of cities and travel, and it has a fundamental role in exploring planning policies. But it is now only one element in the full range of factors which have to be considered. In his discussion of the 'American city theoretical', Hall (1989) concluded that planning theory had moved away from what planners actually do and the kinds of issues they confront in their working lives. The impasse has to be resolved. Some theorists have moved away from urban planning to political economy (for example Scott (1969) and Castells (1983)), whilst others (for example Markusan (1985), Bluestone and Harrison (1982)) have proposed linking political economy approaches with active regional policy prescriptions for older industrial regions. Another group (for example Friedmann (1973) and Forester (1980)) have maintained the more traditional link between theory and action. Apart from trying to reconcile theory with practice and formulating an appropriate role for planning, the grass roots movement is returning to hands-on planning to cover economic redevelopment, project finance and affordable housing.

The dilemma facing transport planners is that even if it is accepted that the methods available are no longer appropriate, that the issues to be tackled are more complex, and that the policy context has fundamentally changed over the last decade, the problems still remain and important decisions, particularly on investment, have to be made.

The 1990s will be a decade of huge investment in the new transport infrastructure, to cover road, rail and air. The scale of that investment will be unprecedented as extensive renewal is required and as networks become fully international. These transport investments will be linked in with other projects on a grand scale. Included here are the Paris Eurodisney, London's Eastern Corridor, a new high speed rail network in Europe, and East-West development in Europe. Investment will be a combination of private and public capital with such issues as investment appraisal, joint venture funding, and ecological impacts all becoming key factors in the analysis. The traditional single-centred or poly-nucleated city will also change as technologically led deconcentration takes place (table 6.1), and as the work ethos is replaced by one based on increased leisure time and quality of life. These fundamental changes in society will be discussed in greater detail in Chapter 7.

From the viewpoint of transport planning analysis (and of urban planning), the need for a renaissance at both the strategic and local levels is required. Three basic reasons can be argued for such a renaissance. Even after a pragmatically led decade where political and ideological concerns were high, some form of strategic framework is necessary. But it is likely to be very different to that advocated in the 1970s and 1980s which was based on the production of large-scale proposals to meet expected shortfalls in capacity. Various forms of strategic and regional guidance would be maintained, irrespective of whether county councils were abolished in Britain and replaced by regional assemblies. Within the existing or modified framework for planning, transport planning would perhaps adopt a more flexible form of contingency approach (Alexander, 1984), where empirical research findings are linked more closely to normative prescriptions. Such an approach would allow new ideas, research methodologies, and evidence from research in Europe and elsewhere to be fed into a strategic framework.

Whether investment is public or private sector funded, some stability is required in the decision-making framework so that decisions can be made with some certainty, otherwise only low risk strategies will be adopted. This means that in transport, only those projects which are underwritten by the government would actually proceed. If the private sector is to be fully involved in the new generation of investment, then a planning framework is required and analysis methods are also essential, together with the ability to learn from past experience and from overseas experience.

The second argument is that predicted increases in travel demand over the next three decades cannot be met. The expectation that new capacity could be provided, either through new construction or through traffic management, to meet that growth has now been firmly rejected. Even if it were possible, it has been seen as undesirable. This new realism (Goodwin *et al.*, 1991) has focused the minds of all those involved in transport planning. The principle of the new consensus is this recognition that there is no possibility of increasing road supply to a level which approaches the forecast increases in traffic.

It logically follows that:

- whatever road construction policy is followed, the amount of traffic per unit of road will increase, not reduce; ie all available road construction policies only differ in the speed at which congestion gets worse, either in its intensity or in spread.

- therefore demand management will force itself to centre stage as the essential feature of future transport strategy, independently of ideological or political stance. (Goodwin *et al.*, 1991, p. 111).

For the first time, even though transport has been a reasonably high priority on the political agenda, there is now a consensus as to what should be done politically and professionally. This means that radical action could take place provided that public support can be obtained, and the resulting requirements for analysis could break the traditional mould of transport planning.

The third issue is that of quality of life. In the past, the main concern has been over increasing the quantity of travel, the acquisition of a car and the notion of the freedom to use that car. As affluence increases, other factors related to the quality and environmental responsibility become important, and values change. The technological revolution now taking place allows such a transition. From the viewpoint of transport planning, the imperative to predict the growth in demand and the overriding importance of economic factors in assessment now becomes less dominant. Methods would be developed to measure quality of life, social impacts, and the environmental/ecological costs of transport. This change in societal priorities should mark a move away from the necessity to quantify everything and to ignore or devalue those factors which cannot be quantified. This switch would result in the replacement of the 'Anglo Saxon' approach to analysis with its narrow conceptual framework and its allocation of a market role to transport with efficiency and productivity being the prime objectives. In this regime, intervention only takes place to correct market failures or to make adjustments for social reasons. Perhaps a more European approach to transport planning will be adopted which will assign transport the status of an intermediate activity that requires direct control to achieve wider social, industrial, regional and national objectives. To this list would be added quality of life and international objectives.

The optimistic scenario presented for the 1990s is that transport planning will undergo a renaissance, in a modified form as the requirements of strategic planning, the new realism, and the quality of life imperatives dictate. At last, the TPM appropriate for application to large cities in the 1960s and 1970s has been retired. Although mainly dormant in the 1980s, the systems approach was still influencing thinking, but events in politics, in the growth of demand, in new research directions, and in public attitudes have all resulted in a fundamental change. Such radical shifts have occurred before, but never has such a consensus existed.

*Chapter 7*

# Transport Agenda 21:
# The Way Forward

Much of the evidence presented in the previous Chapters has been critical of transport analysis, arguing that it only takes a narrow single sector perspective of the problem. Transport has not been seen as an integral part of all activities and its role as a means to achieve access. In the next two Chapters the broader issues are presented, first in view of the major policy changes which are likely to emerge over the next decades, and secondly in viewing the role that transport planning can and should take.

Transport is the means to an end and not just an end in itself. As such it should be seen in the context of changes which are taking place more widely in society. The role of transport planning should be to facilitate access to and participation in activities. But it should also ensure that all groups within society benefit, that broader objectives (for example on cost and the environment) are met, and that the quality of life for all is maintained and improved. Some of these objectives are likely to be in conflict with each other, and in these cases planners should assess all the evidence before advising decision-makers what to do. One clear lesson from the 'market experiment' in the 1980s has been that the market works well in certain situations, but there is still a need for an overall strategic framework within which the market can operate. It is in the definition of this framework that planning has an important role to play.

Transport planning has tended to limit itself to a narrow base, looking at problems from the transport perspective and presenting solutions only in terms of transport options. Even within transport itself, the range of options has been mainly restricted to pricing and physical measures. Other possibilities are either defined as exogenous to the process (for example demographic change and the structure of the economy) or as planning options (for example land-use strategy). Increasingly, problems, analysis and solutions must be examined holistically as all these questions are part of the same process.

This Chapter presents some of the major changes taking place in society which are likely significantly to affect the demand for travel into the next century. Three main issues are identified, which relate to demographic change, technological change, and the infrastructure requirements. It is argued that the Agenda 21 for transport planning must address all three. Each of the main themes has been placed in an explicitly European framework as transport planning within the UK is likely to share a much greater commonality with approaches adopted elsewhere as greater interchange of ideas and people takes place within the European Community (EC). Decisions on and funding of major transport investments will take place at the EC level as greater cohesion and integration within the EC is achieved, and as the increasing costs of infrastructure mean that loans and grants will come from the EC and the European Investment Bank. More travel will be international with freedom of movement, the ability to work in any EC country, and the expected growth in affluence.

## Demographic Change

Some of the broad trends in population have been outlined in Chapter 1. In this section the main empirical evidence from the last 20 years is summarized and two themes important for transport planning are developed. These are the 'motorization' revolution and the 'grey' revolution. Taking the empirical evidence from the last 20 years it seems that certain conclusions can be drawn concerning the changing travel patterns in Great Britain.

- There has been a steady growth in travel with journeys made increasing by 18 per cent (from 11.2 (1965) to 13.2 (1985–86) journeys per person per week).

- Within these totals the journeys made for work related activities (that is work journeys, journeys made in course of work, and education journeys) have declined marginally (from 4.4 (1975–76) to 4.2 (1985–86) journeys per person per week).

- The greatest increase in activities has been in the leisure activities which include social and entertainment purposes as well as holidays and daytrips (from 3.8 (1975–76) to 4.3 (1985–86) journeys per person per week). These activities now account for more travel than work related activities.

- There are great similarities between the aggregate journey patterns of men and women (16–59 years) and the elderly in car-owning households for non-work related activities (table 7.1).

- There have been significant increases in travel distances from 1965 to 1985–86 (+41.5 per cent) and this increase is apparent across all activities and all sectors of the population. The greatest increase seems to be in the

TABLE 7.1 *Non-work travel per person per week in Great Britain, 1985–86.*

| | Distance Travelled (miles) | Journeys | Journey Length (miles) |
|---|---|---|---|
| Children | 47.6 | 7.0 | 6.84 |
| Men 16–59 | 78.4 | 10.0 | 7.84 |
| Women 16–59 | 74.9 | 10.3 | 7.27 |
| Elderly >60 | 51.4 | 7.7 | 6.68 |
| Elderly drivers | 87.4 | 11.4 | 7.64 |
| Elderly in car owning households | 75.8 | 10.5 | 7.23 |
| All People | 65.0 | 9.0 | 7.22 |

*Note*: Elderly drivers account for 32 per cent of all people in the elderly category and include main drivers and other drivers in car-owning households. These figures are the weighted average for the two car driver groups. If non-drivers in car-owning households are included, this covers 48 per cent of all people in the elderly category and the weighted figures for all elderly in car owning households can be calculated.

*Source: National Travel Survey 1985 86.*

elderly age group (about +76 per cent, but it should be noted that the age groupings are slightly different between the two travel surveys). The only below average increase has come in men between 16 and 64 years (+37 per cent). These percentages distort the absolute increases in mileage where the male dominance is still apparent and increasing (table 7.2).

Much of this growth in mobility has been made possible by the growth in car ownership, which is apparent in all OECD countries where the close link between car ownership and GDP per head can be observed. Britain falls around the centre of the distribution and even now may have considerable further potential for growth in car ownership levels. Mobility patterns are strongly correlated with car ownership patterns (table 7.3). The distance travelled by people in non-car-owning households has remained stable over

TABLE 7.2. *Changes in distance travelled per person per week in Great Britain, 1965 to 1985–86.*

| | 1965 Mileage | 1985–86 Mileage | Absolute Increase | Percentage Increase |
|---|---|---|---|---|
| Children | 39.2 | 56.6 | +17.4 | +44.3 |
| Men 16–59 | 118.2 | 163.2 | +44.7 | +37.7 |
| Women 16–59 | 64.1 | 100.2 | +36.1 | +56.3 |
| Elderly | 33.0 | 58.2 | +25.2 | +76.4 |
| All people | 70.3 | 99.5 | +29.2 | +41.5 |

*Note*: Some of the age groups have changed and they are not strictly comparable – the figures are only indicative.

*Sources: National Travel Surveys.*

TABLE 7.3. *Travel distance per person per week by car ownership in Great Britain.*

| | People in Households with | | | *All People* |
|---|---|---|---|---|
| | *No Cars* miles % | *One Car* miles % | *Two or More Cars* miles % | *miles* |
| 1965 | 42.0 (59) | 96.0 (36) | 132.0 (5) | 70.1 |
| 1972/73 | 40.6 (48) | 103.8 (44) | 134.5 (8) | 82.0 |
| 1975/76 | 40.8 (44) | 101.8 (45) | 144.2 (11) | 85.9 |
| 1978/79 | 48.6 (42) | 110.0 (45) | 150.5 (13) | 92.6 |
| 1985/86 | 40.6 (38) | 104.5 (45) | 162.7 (17) | 99.5 |

*Source: National Travel Survey 1985–86.*

the last 20 years at about 40 miles per person per week. In households with one car there has been a small increase of about 10 per cent, but in households with more than one car the increase is over 23 per cent with each person on average travelling some 160 miles per week. These changes in distances have been compounded by the growth in car ownership with non-car-owning households declining from 59 per cent to 38 per cent and single-car-owning households stabilizing at 45 per cent. The growth is in households with more than one car and these households also have had the greatest increases in weekly travel distances. In conclusion it seems that the summary points made on changes in travel patterns for the different demographic groups are compounded by car ownership changes. These two themes are developed in the remainder of this section.

THE 'MOTORIZATION' REVOLUTION

Traditional analysis assumes that trip rates for particular age cohorts remain stable over time and that as population ages it adopts the travel characteristics currently undertaken by that group. For example, a person in an age group 40–50 today will adopt the current travel characteristics of the age group 60–70 in 20 years time. The counter argument put forward here is that all age groups, but particularly the elderly, have experienced a motorization effect which will fundamentally affect their habits and expectations. The 1990s is the first decade where the elderly (>60) have experienced full motorization as they were the first generation of mass car ownership. They were in their thirties in the 1960s. Consequently, it is unrealistic to expect these people to take on the travel characteristics of those individuals who have never experienced full motorization. These issues as they relate to the elderly are covered fully in the next section, but in this section the related issue of car ownership is discussed. The corollary to the argument for dynamic approaches to analysis is that individuals will try to maintain the car and the ability to drive as long as possible. Hence it becomes important to discuss car ownership and car use, and the likely effects that these two critical factors will have on social behaviour.

A demographic analysis of car ownership and use patterns takes as its starting point the growth in licence holdership and car ownership for the different age groups for men and women. Evidence is taken from the USA, Norway and Britain on growth trends over the recent past and the likely saturation levels. As such it differs from most car ownership forecasts where growth is considered to relate closely to increases in real incomes or GDP and the real changes in the costs of using the car (Department of Transport, 1989*b*). These econometric forecasts complement demographic trends which examine the propensity of different cohorts to have a driving licence, to own and to use a car. As Goodwin comments, in 1990 we were about half way from zero car ownership to saturation in 2025. But most of this growth has taken place in one working life (Goodwin, 1990).

In the USA, 59 per cent of households had at least one car in 1950 and by 1987 this figure had risen to 87 per cent. The number of households with two or more cars increased from 7 per cent to 52 per cent over the same period. The current number of cars per 1000 population is 565, rapidly approaching the assumed saturation level of 650. The saturation level is based on the assumption that 90 per cent of those between 17 and 74 years of age will want to have a driving licence and a car. There does seem to be empirical evidence in the USA to support this argument (Greene, 1987). All men in the 25–69 age group now have a driving licence and there has been little change between 1969 and 1983. In the under 25 year age group about 80 per cent of those eligible to have a driving licence do have one, and it is only in the elderly age group (>70) that there is some fall off in licence holdership. But even here, in 1983, over 83 per cent of all men over 70 years of age still have a driving licence. For women the picture is somewhat different and there has been a huge catching up process between 1969 and 1983. In the 25–40 age group over 90 per cent of women have a driving licence, and this level falls in the 41–60 age group to 85 per cent. In the under 25 age group the level is 70 per cent. Of the women over 70 years old only 40 per cent have driving licences (FHWA, 1984). There does seem to be some convergence between men and women on driving licences, but the miles driven by each do not converge as they are both increasing by about 2 per cent per annum (Greene, 1987).

This growth means a strong divergence as men have historically driven more than twice the distance that women have in the USA. In 1983, the average annual distance driven per licensed male driver in the USA was 22,350 kilometres, whilst the corresponding figure for females was 10,200 kilometres (US Department of Transportation, 1986). Estimates can be obtained of travel demand given the growth in travel distance and the increase in licensed drivers. It would seem that in the USA, saturation levels of drivers and cars should be reached by the year 2000, some 25 years before Europe. However, it also seems true that the desire to own a car will not end here. There is a strong trend towards multiple car ownership with the greatest growth occurring in households owning three or more vehicles. Prevedouros and Schofer (1989) put forward an argument that the value of owning and using cars exceeds the value of the transport produced.

> the automobile may be viewed as an office, a storage unit, a home away from home, a means and place of recreation, and a social instrument of increasing significance; people increasingly define themselves by the number and types of automobiles they own. As a result, automobile mobility may have social status value beyond that reflected in travel forecasting models.

In Europe the picture is somewhat different. In a recent paper, Vibe (1991) takes data from three surveys in the Oslo region for men and women with driving licences and their own car. He concludes that the level of licence holdership and car ownership for men is stabilizing across all age groups at about 85 per cent with a lower figure for the 25–34 age group of 80 per cent, and for the youngest age group (18–24) a level of about 55 per cent. For women, a completely different conclusion is drawn. Here he suggests that there is still considerable potential for growth with a saturation level of between 65 and 70 per cent. Licence holdership for all women in all age groups has increased by about 20 per cent over the three surveys. For all people the level of 75 per cent is suggested as being realistic for the population between 18 and 80 years by the year 2000. The present level of 64 per cent and nearly 400 cars per 1000 population will increase to 470 cars per 1000 population. Vibe (1991) suggests that this is a saturation level in Oslo which is higher than the current levels in the former Federal Republic of Germany (461 per 1000 in 1988) but still far below the US level (565 per 1000). The crucial unknown factor in these calculations is the saturation level of licence holdership for women; but even if it reached the 85 per cent level assumed for men, this would still only give an upper limit of 531 cars per 1000 population.

In Britain, the numbers of men and women with driving licences are remarkably similar to those for Oslo in 1977 for both men and women. This similarity is continued to 1985–86 for men, when some 86 per cent in the 30–50 age group had licences and over 80 percent in the 50–60 age group were still drivers. The two age groups (20–29 and 60–69) had over 70 per cent licence holdership. If that comparability extends to the present then one would expect that 85 per cent of British men between 30 and 75 years would have driving licences. There are no figures in Britain to corroborate this. For women there are differences between Oslo and Britain with higher licence holdership figures in Norway in the youngest and eldest age groups, but lower across the middle age groups where British women have licence holdership levels of 62 per cent. This might suggest that the saturation level for women in Britain is higher than that for women in Oslo.

If the same types of assumptions are used in Britain, with 85 per cent of men 20–70 and 55 per cent of the youngest and eldest age groups having driving licences and with the corresponding figures for women being 75 and 50 per cent, the levels of car ownership will reach 436 cars per 1000 persons shortly after the turn of the century. If the 85 per cent figure across all age groups for both men and women (17–80 years) is used, the resulting figure is 494 cars per 1000 persons, well below the lowest figure in the official forecasts for 2025.

The National Road Traffic Forecasts (Department of Transport, 1989*a*) suggest that traffic in Britain will double between 1988 and 2025, with car ownership levels increasing from 331 cars per 1000 population to between 529 and 608 cars per 1000 population. About 46 per cent of the population between 17 and 74 years had a car in 1988 and this proportion will increase to between 73 and 84 per cent. This means that the saturation level, given the assumptions stated previously, will very nearly be reached by 2025, and the levels of car ownership in Britain will be close to those found at present in the USA, and significantly higher than those suggested in Oslo. It also means that women are just as likely to drive and own a car as men and that the elderly will retain the use of the car as long as possible.

The important connections between car ownership and demographic factors can be summarized:

- Over the next 30 years there will be significant increases in car drivers and in the numbers of cars (averaging between 60 and 80 per cent) in many European countries to around 550 cars per 1000 population or similar to the current USA levels of car ownership.

- Much of this growth will take place in households where there is already one car.

- Men already have high rates of licence holdership and this will remain stable at between 85 and 90 per cent of those aged 20–70. The greatest growth in licence holdership among men will be in the youngest age group (17–20) and those in the elderly age group (>70).

- For women the growth in licence holdership is likely to be significant across all age groups to between 70 and 80 per cent. The greater access of women to cars as drivers is likely to be a major source of new travel. Although some of these new activities will be work related as women increase their participation in the labour force, much will reflect their multivariate and complex pattern of activities which involve accompanying others, personal business, shopping activities as well as social and recreational activities.

- The expectations of young people will be raised, with many aspiring to car ownership. Again, given rises in income levels, it is likely that as soon as they can drive, young people will want to do so and here again one can expect a significant increase in travel demand.

- The elderly age group is the third and perhaps the most important growth area for travel demand. This group has traditionally had low levels of car ownership and licence holdership, but over the next 30 years one can anticipate that this will change dramatically. Many of those who will be elderly in 2025 are at present among the highest car ownership and licence holdership groups, and they are also the most mobile. This is the 'grey' revolution to which we now turn.

## THE 'GREY' REVOLUTION

The 1990s could be the decade when grey power becomes a major factor in determining levels of travel demand. It is widely assumed that the elderly have low levels of mobility and this is illustrated in table 7.4 where the elderly have not only low levels of mobility but also a high dependence on the bus. However, it is important to realize that the elderly may not have the same travel demand patterns in the future as the corresponding group has at present. A systematic study carried out in the USA (Kostyniuk and Kitamura, 1987) has found that a 'motorization effect' has to be linked with an ageing effect. Increases in mobility have taken place across all age groups, but the increases have been much greater for men than for women (1963–1974). Both cohort and time effects at given levels of motorization influence travel patterns and so future elderly cohorts will behave differently from those of today. Studies of the elderly as transport disadvantaged have often assumed that they have low income, low car availability and some physical disability. But these assumptions are increasingly being violated and it is essential to include a dynamic element for the changing experiences and expectations of the elderly.

In addition to this increasing tendency to keep the car as long as possible, there are also other factors which lead to the conclusion that one major growth sector in travel demand will be from the elderly. Over the next 30 years significant growth will occur in the numbers of elderly. This growth is partly due to demographic factors and the increase in life expectancy, and partly due to definitional factors with the tendency to retire earlier. It is estimated that in Western Europe the proportion of persons over 65 years will increase from 13 per cent in 1985 to more than 20 per cent in 2020. For OECD countries the

TABLE 7.4. *Journeys per week by age 1985–86 in Great Britain.*

|  | Car Journeys | Bus Journeys | All Journeys |
|---|---|---|---|
| Children | 5.6 | 1.8 | 9.8 |
| Men (16–59) | 13.7 | 1.3 | 18.0 |
| Women (16–59) | 10.1 | 1.9 | 14.5 |
| Elderly | 5.4 | 1.8 | 8.6 |

*Source: National Travel Survey 1985–86.*

number will increase by 50 per cent from 98 million in 1990 to 147 million in 2020, and peaking in 2040 with 175 million people over the age of 65. The growth is particularly significant in the age group >80 which is projected to triple in the next 50 years. Taking the decline in fertility rates and the growth in the elderly population means that not only does the absolute number of elderly double, but the proportion of the elderly increases due to the relative decline in the younger population (OECD, 1989).

Within the elderly population there will be considerable variation between those who are affluent and highly mobile car drivers, and those who are disabled or on low or fixed levels of income. It may be appropriate to subdivide the elderly age group into the mobile age of personal fulfilment (60–80 years: Laslett's 3rd Age or 55–80 years if early retirement continues (table 1.2)) which is likely to remain stable over the next decade, and those who are fully retired and dependent (over 80 years: Laslett's 4th Age) which is the major growth sector in the next 30 years. Within each of these two groups there is also likely to be considerable variation in travel demand patterns, as it seems that age is a key determinant of mobility. As senility and age-related disabilities occur, significant changes in activities and lifestyles do take place.

The personal fulfilment group have ended the complex responsibilities of earning a living and raising a family, they are reasonably affluent, and so have both money and time to spend on leisure based activities. Many of these activities will involve travel, and their purpose may be to visit friends or relatives or to achieve a life long ambition. Alternatively, the mobile early retired groups may also be involved in a wide range of social and voluntary activities where their time, skill and knowledge can be given to an activity and where no payment is received.

In Britain there has been a steady growth in real incomes over the last decade and this has been augmented by increased levels of inherited wealth. As a result of unprecedented increases in levels of house prices, large amounts of capital are now available for spending or passing onto one's children. Levels of home ownership have been traditionally high in Britain, and with the right to buy the council house in which one was living at lower than market prices, some 70 per cent of people now own their homes. As there is no tax to pay on the sale of one's own home, housing forms the main capital asset of many families. With retirement, people often decide to move to a smaller house or to an area where housing is cheaper. In 1990 the inherited wealth from the sale of property amounted to £8 billion. Tax free capital is released so that more consumer expenditure can take place. Similarly, people are borrowing against actual or expected rises in house prices, and this wealth together with income growth and available credit has fuelled the increases in consumer spending which was a feature of the late 1980s. It is estimated by the Mortgage Corporation (Hamnett *et al.*, 1991) that the amount of capital that will be released over the next 10 years by this mechanism will be a massive £29 billion.

Although house prices are at present falling (1989–1993) as the gearing between prices and wages is gradually readjusted, the underlying increases in value are likely to continue, albeit at a much more modest level.

It will only be when people reach the age of 80 that full retirement and dependency takes place. This group of people, which accounts for about 21 per cent of the elderly, will require special facilities and transport services which can accommodate their particular requirements, for example to be wheel-chair accessible or to have a person to accompany them. This group will not be able to drive a conventional car and so will require public transport services or taxis or chauffeur-driven private cars. The voluntary sector organizations provide many of these services.

One possible development might be to design a vehicle specifically for the elderly to give them some degree of independence, perhaps similar to the battery operated tricycles which are already available with voice activated controls to ease the physical requirements of driving. Special routes could be provided for these low performance vehicles which could be used by the elderly and perhaps children.

The empirical evidence (Banister, 1992c) seems to indicate that for given car availability conditions there is a certain stability in non-work trip making by age. If this simplifying assumption holds for the next decade, calculations for the number of journeys made and distances travelled for the elderly can be made. Depending on the strength of the assumptions used, the growth in journeys made by the elderly will increase between 23 and 34 per cent with journey distances per person per week increasing by between 30 and 50 per cent. The modal split will also change with an increase in trips as car driver and car passenger from 60.5 per cent (1985–86) to about 75 per cent (2001). The elderly will no longer be seen as a disadvantaged group in terms of their travel patterns as these will now fully reflect non-work activities of those currently in the 45–60 year age group. Use of public transport by the elderly, principally bus services, will also decline by between 14 and 24 per cent over the fifteen year period. The implications for public transport operators are severe as one of the major user groups for bus services has traditionally been the elderly, and their loyalty has been assumed captive. This assumption may no longer hold.

Public transport operators may need to start planning for an elderly population which will want a much greater quality of service provision with convenient and comfortable services that are easy to get on and off. Other users may require door-to-door services with an escort facility. The disadvantaged and physically frail are one group who will continue to be dependent on public transport. Their quality of life needs careful consideration as the effects of ageing seem to progress rapidly after the age of 80, and these people are often living on their own. As noted earlier, some 84 per cent of those living on their own, who are over 60 years, have no car and these account for 16 per cent of all households. It will not be cheap to provide a quality service for these people who have traditionally had low levels of mobility, and much has depended on the voluntary sector to provide back-up to the statutory sector (Banister and Norton, 1988). These services are unlikely to be cost effective or to make a profit, and so the allocation of subsidy and the opportunity costs of that subsidy will have to be considered. Evaluation will have to take on social and equity criteria as well as conventional efficiency criteria.

It seems that two types of public transport services are required for the elderly. On the one hand, there would be a quality service provided by the operator at a premium price with no subsidy. This service would accommodate the day-to-day requirements of the mobile elderly without access to a car, and would also be a leisure mode to take them on day trips and to visit friends and relatives. On the other hand, there would be a special service for those with low levels of mobility which would provide door-to-door access with wheel chair and escort facilities. This subsidized service would enable the physically frail and elderly to travel to shops, to day care centres, to the local hospital and to friends. This is the crisis facing public transport. Traditional captive markets are diminishing and new types of services are required for the elderly, but also for the young and for women. People will choose to travel by public transport and not because they have no other alternative.

OTHER DEMOGRAPHIC AND RELATED CHANGES

In addition to the changing demographic structure of the population, other factors are also likely to affect behaviour and the demand for public transport.

1.  *Returning Young Adults* usually result in an increase in the number of cars available in a household and an increase in the trip generation rates. These highly mobile young people are returning home after a period of education and training before they get a permanent job and set up home. This phenomenon may increase in times of high unemployment or high property prices, both of which make independence uncertain and expensive.

2.  *Female Participation Rates* in the labour force have increased, particularly for part time work and for those returning to work after raising a family. In Britain it is expected that 90 per cent of the one million increase in the labour force (1988–2000) will be women. The activity rates for men have remained constant at 73 per cent, whilst that for women has increased by 12 per cent over the last 10 years to nearly 50 per cent. The greatest growth in female participation rates has been and is likely to continue to take place in the sunbelt countries of the Mediterranean.

3.  The decline in average *Household Size* across all developed countries by about 20 per cent over the last 30 years reflects lower fertility rates, but

also the breakdown of the conventional family unit. Many children live in single parent families, divorce rates have increased, more people now cohabit, and there are a larger number of births outside marriage. These trends have an impact on the housing market with an increase in demand for smaller units, which would be located either in the city centres through subdivision of existing properties or in the suburbs in purpose built units. In each case the ratio of car parking spaces to homes will be near unity and this may create parking problems, particularly in city centre locations. If the location is in the suburbs, then the numbers of trips generated and the length of those trips is likely to be greater.

4.   With increased *Suburbanization and Greater Complexity of Work Journeys*, the pattern of daily movements is becoming more varied both spatially and temporally. Commuting patterns have become more complex with cross commuting becoming more important than commuting to city centres. Households with an established residential base are likely to meet their career needs by longer distance commuting. In a tight labour market (such as that in the South of England), there is often more than one person employed in the household and these complicated travel patterns emerge as the transport system has to accommodate to this change. With high interest rates and low levels of residential mobility, it is again the transport system that has to respond as people cannot move home. In times of economic recession, job and housing mobility are also reduced; travel is curtailed and non-essential activities are limited. The cost of transport may present a major constraint on job search areas as the likelihood of being successful has to be balanced against the travel costs involved.

5.   Increased *Leisure Time* and the shortening of the working week may also affect the demand for travel. Trips taken for social and recreational activities have increased by over 28 per cent between 1975–76 and 1985–86, and leisure accounted for 16 per cent of all household expenditure in Britain in 1989. Most of this increase can be explained by the growth in social activities as holidays and daytrips have remained constant over time. Shorter local trips by walk and bicycle have been replaced by longer trips overseas and by car based trips.

6.   The greatest unknown factor is *Migration* patterns in Europe with the opening up of the Single European Market and the breaking down of barriers between East and West. Although natural population growth may be very low, significant changes in population could be brought about by internal migration as labour seeks employment and as industry moves to locations where labour is available, and where other factors such as land availability, environment, and communications are positive.

All these factors suggest that the demand for travel will continue to increase and that the present pattern of longer journeys will also be maintained. This means that walk, bicycle and bus trips may continue to decline as these journeys are replaced by the car. Superimposed on all these demographic factors is an underlying change in society brought about by the car, namely the 'motorization' effect or what the French have called 'irreversible behaviour'. Low motorization is not an inherent characteristic of particular groups within the population, but they have been impacted by it because of the cohort effect (Kostyniuk and Kitamura, 1987).

With the growing number of drivers it is likely that societal pressures will be brought to bear on people deemed to be antisocial in their driving behaviour. Excessive speed will not be acceptable as the majority of new drivers may favour lower speeds. Safety and environmental concerns over the use of resources and increasing levels of pollution may result in pressure being exerted on manufacturers to produce 'green' cars. These vehicles would only be capable of being driven at low speeds and so would prove attractive to the elderly and other people, and the age of acquisition of a driving licence could be reduced to allow the young to acquire this type of vehicle. In Japan 90 per cent of elderly drivers (60–69) want to continue driving, and 80 per cent of those over 70 feel the same way even though 30 per cent are under family pressure to stop driving (Orfeuil, 1991). In the USA, there is a concern over the increasing numbers of road accidents involving elderly drivers. This trend is likely to continue as more elderly people continue to drive until they are forbidden, either through family pressure or through legislation.

There must be a market for a safe low-speed environmentally benign vehicle which would form an intermediate stage before the acquisition of one's first car (for the 13–16 age group) and before the complete loss of personal mobility by the elderly (for the 70–80 age group). These two groups account for nearly 13 per cent of the population. Individuals have concerns over driving at night. Their own security is increasingly important for all modes of transport. Together, these factors may explain why some people will not go out or drive at certain times, even if the car is available. Safety and security in travel are two important deterrents to road users and non-road users.

The alternative to providing more or different forms of transport to accommodate the expected changes in demographic structure must be through positive planning policies which allow mixed development and the provision of local facilities to minimize the need for car travel. Car dependence can be reduced by the use of the new quality public transport for local travel together with the 'green' car and the bicycle. At present lifecycle changes influence both the mobility patterns and the location decisions of households (table 7.5).

The question here is what is the most appropriate use in residential terms of the city centre. Often it is too expensive for young couples or workers with children to locate there, and it is the institutions, families of adults and the

TABLE 7.5. *Lifecycle effects on travel and location.*

| Lifecycle Group | Travel Patterns | Location |
|---|---|---|
| Individuals or couples without children, often all working | High level of mobility with maximum levels of car ownership | City centre |
| Families with young children | Age of youngest child influences adults' travel patterns. Wage earner keeps former pattern but other partner develops short day time trips and complex escort trips with older children | Suburbs |
| Families of adults but no children, often all working | Children independent and separate patterns develop for all households with maximum car ownership | City centre |
| Elderly and retired | Substantial opportunities for non-work travel. Mobility decreasing with age | Some stay in city centre and others move to suburbs or rural areas |

elderly who have city centre property. The other lifecycle groups are forced to live in the suburbs or out of the cities and to undertake long-distance commuting. The work journey is accepted as a cost to be paid. Allowing workers to live near their workplace would reduce travel distances, but this simplifies the reality of the location decision.

The last two decades of mobility growth seem likely to continue for the next two decades as important groups within the population acquire driving licences and cars. Unprecedented growth in car use by the elderly and women seems certain. If the car manufacturers can produce a vehicle that is environmentally attractive and efficient to use at low speed in towns, then the young teenagers and the more dependent elderly will also join the motorized culture. The future for public transport is less clear and positive action is required to increase its attractiveness to all potential users as it can no longer depend on the diminishing captive user group. Demographic factors together with economic, social and technological changes mean that the relentless increase in demand for travel will continue. The breaking down of political boundaries, the increase in international travel, and the expected real growth in income levels all suggest that we may be close to the final gridlock when not only our cities come to a standstill, but the whole of Europe grinds to a halt.

## Technological Change

The main source of profit and power in the late twentieth century is knowledge and information, and conflicts are likely to occur over the distribution of and access to that knowledge. The control of knowledge is power. Even money is becoming less important and tangible as transactions are carried out electronically, only to be seen in a symbolic form on a screen. This global view of the shifts in power presented by Toffler (1991) will transcend all activities at all levels and will create a radically different society. It will also be instrumental in the current road transport revolution with the possibility of wired cities, wired highways and wired cars. The availability of knowledge and information will radically change road transport in at least four related ways.

In production and distribution processes, the production line methods introduced by Henry Ford at his Highland Park assembly plant in 1913 are now being extended and replaced. The conveyor belt now extends beyond the manufacturing system to the distribution system. Technology and information allows a complete service from the assembly of materials through the production of the car to the testing and distribution processes, and the delivery to the final consumer. These concepts do not just apply to the manufacture and distribution of vehicles, but to all commodities. Freight distribution systems have been restructured on regional and metropolitan warehousing depots, often at accessible motorway intersections. Road transport informatics (RTI) impact on all parts of freight transport operations as well as location decisions (figure 7.1). With the trend in Europe towards longer distance trucking and

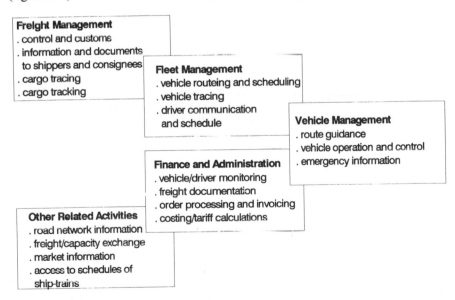

FIGURE 7.1 Information Technology Applications in Freight Transport. (*Source*: Pangalos (1989))

the increased use of multimodal combinations of vehicles and carriers, integrated approaches to freight transport are essential to ensure the optimal use of information and the new flexibility in both production and distribution processes. The potential is available to develop a Europe-wide integrated freight transport network, but old barriers still remain, namely who should pay the costs of pollution, the increased resource costs caused by the growth in international road freight, and the compensation of individual member countries for transit traffic – 'the territoriality issue'.

Particular transit countries within the EC suffer from traffic generated at peripheral countries. These costs incurred by transit countries should be the responsibility of the road users and it is likely that a charge will be levied according to the distance travelled and the type of vehicle. However, such a charge is against the notion of a Single European Market and the liberalization of the transport industry. RTI will be used to levy such a charge through automatic debiting and an intelligent tachometer could be used to record the trip history so that there would be no need to interrupt the journey (Hepworth and Ducatel, 1992). These user taxes are likely to form only a small part of total production and distribution costs, and so the impact on the location and competitiveness of industry will be limited.

The movement of people provides the second opportunity for RTI with the belief that technology can help delay the inevitable gridlock when the city comes to a complete stop through congestion. Traffic management schemes have been very effective in squeezing more capacity out of a given road network, and the expectation here is that technology through intelligent highways and smart cars can continue that process. Increased flexibility in work and leisure patterns together with the possibility of telecommuting have all provided the opportunity for change. Again, it should be noted that information and knowledge have both been instrumental in creating the conditions for this opportunity. However, each revolution in the past has resulted in increases in travel and average trip lengths, and there is no reason to expect a change as a result of the current revolution. As extra capacity is created, demand increases.

Road users will be affected in three different ways:

- Information services to the traveller which will allow decisions to be made on the basis of the best real time information. These services would apply equally to public transport services and to route guidance information given to the car driver.

- Control systems within the vehicle. By the year 2000, it is estimated that 10–15 per cent of the costs of new cars will relate to RTI services (Lex Motoring, 1992).

- Control over the transport network, including demand management and traffic control systems.

RTI will also have considerable commercial applications through more efficient management of freight companies and public transport services.

The potential impact of RTI on growth in car ownership and use should not be underestimated. In all European countries over the last 10 years there have been increases in car ownership, increases in trip lengths and increases in the numbers of trips made (table 1.3). RTI may again cause similar increases in personal mobility, but its potential is much wider. Through the use of control systems it may be possible to produce vehicles which are suitable for both the elderly population and those who are too young to drive (see page 171).

The infrastructure is the key to an integrated Europe, and much of the existing transport infrastructure is over 100 years old (see pages 184–191). A significant part of the motorway system is now over 50 years old. Very substantial investment is required to replace existing roads and construct new ones, and these links will help integrate peripheral areas as well as open up new markets in the old East European countries. Road investment will be complemented by the new European high speed rail network and telecommunications networks, including the new Value Added Networks (VANs) and the Local Area Networks (LANs). It is this combination of networks which will facilitate the most fundamental changes brought about by knowledge advances and information technology. These include:

• logistics planning;

• electronic data interchange;

• electronic route guidance;

• emergency transport planning;

• information systems;

• databases for environmental monitoring.

As a consequence, the spatial imperative will no longer apply as cities will become much looser spatial organizations because the costs of urban centrality and high land prices will be balanced against the benefits of dispersal. The movement out of cities will continue with only front office functions remaining. Growth will be concentrated in corridors of good communications and at peripheral urban locations where it is cost effective to link in with both the transport and the information networks. Peripheral areas may still remain isolated and separate from the new infrastructure as access costs and capacity requirements may make the installation costs of the new networks uneconomic and the costs of using the system too high.

The most attractive locations in Europe will be those where the transport and information networks link in with other factors such as a skilled labour force, a high quality environment and the availability of low cost land. Interchanges may provide particularly suitable locations for logistical platforms. International airports, high-speed train stations, and major motorway intersections could all provide the sites of maximum accessibility which would minimize location and transport costs, and also be on the international information network.

There is no question that very significant changes are taking place on both the political and technological fronts in Europe, and RTI is likely to have a profound effect on road transport at all levels. However, the exact nature and scale of that impact is far from clear, and it seems that implementation of RTI will not be equal across all road transport sectors, but that it will be selective and take a considerable time for the full effects to become apparent.

POSSIBILITIES AND POTENTIALS

The scale of potential RTI impacts has been identified in the introduction and this futuristic vision was apparent in the original programme of DRIVE (Dedicated Road Infrastructure for Vehicle Safety in Europe) set up by the EC. However, more recently the concern has switched to market applications in the European research programme, and with the parallel programmes in Japan and the USA, there is now an unstoppable impetus for the application of RTI to road transport. The DRIVE programme is divided into seven areas of major operational interest, and particular attention in DRIVE II (1992–1994) will be given to the validation of research and development results achieved through pilot projects (table 7.6).

As can be seen from this comprehensive list of projects, the potential range of applications is enormous and such a programme has been justified by the importance allocated to transport congestion, safety and technology in the development of EC research. Transport represents more than 6 per cent of GNP with more than 10 per cent of the average family budget being devoted to transport, and there is a strong expectation that with the growth in car traffic 'bottlenecks will inevitably occur in land infrastructure in Europe' in the 1990s (CEC, 1991, p. 5).

These problems exacerbate the negative effects of road transport on human safety and the environment:

- Every year in the Community around 55,000 people are killed on the roads, 1.7 million are injured and 150,000 permanently handicapped. The financial cost of this is estimated to be more than 50 billion Ecu per year. The social cost in human misery and suffering cannot be measured.

- The cost of traffic in the Community is estimated to be around 500 billion Ecu per year. A substantial part is due to congestion and poor routeing. In France alone in 1988 more than 400,000 km hours were spent in congestion.

TABLE 7.6 EC DRIVE II research programme – the seven areas of operational interest.

| Areas of Major Operational Interest | Applications |
|---|---|
| Demand Management | Area access control, area pricing, zone access control, zone pricing, parking control, parking pricing, road pricing at barriers, booked pricing, special points pricing, law enforcement, fares collection, parking management |
| Traffic and travel information | One way communication to vehicles, dialogue, route information, service information, pre-trip information, traffic information, parking guidance, public transport information, fleet management, weather conditions, road conditions, pollution, traffic flow |
| Integrated urban traffic management Integrated inter-urban traffic management | Coordinated junctions, integrated network control, tidal flow, tunnel control, specific road users, emergency services, traffic calming, parking management, route control, heavy goods vehicle control, traffic monitoring, congestion detection, incident detection, environmental monitoring, enforcement/policing |
| Driver assistance and cooperative driving | Cruise control, anti-collision control, intelligent manoeuvre control, cooperative intersection control, lane merging, medium- range information |
| Freight and fleet management | Dynamic route planning and scheduling, fleet monitoring, vehicle dispatching, intermodal transport planning, transport order processing, consignment monitoring |
| Public transport management | Dynamic data on traffic conditions, information database on public transport schedules and networks, real-time information on vehicle locations, travel demand forecasting, scheduling software, in trip terminals, real-time information linked to traffic control centre, pre-trip information, automatic debiting, fare collection and ticketing, public transport priority, real-time and static interchange information, real-time in-car/in-bus information, park-and-ride linked to route guidance |

Source: EC DRIVE II BATT Consortium (1992).

In the UK the Confederation of British Industry estimates that more than £15 billion (21 billion Ecu) are lost each year in congestion.

• Vehicle emissions contribute significantly towards the total of environmental pollution which is estimated to cost Europe between 5–10 billion Ecu per year.

This is the level of the current challenge facing the EC, and if current trends are continued, the demand for road traffic will increase by a further 34 per cent over the next 10 years. International road freight along the main corridors will increase by between 13 and 15 per cent per year. The DRIVE programme attempts to improve road safety, maximize road transport efficiency and contribute to environmental improvements. It envisages a common European road transport environment in which drivers are better informed and 'intelligent' vehicles communicate and cooperate with the road infrastructure itself (CEC, 1991, p. 5). This Integrated Road Transport Environment (IRTE) will be achieved through pre-competitive and collaborative research and development, the evaluation of systems, the harmonization of European standards and common functional specifications, and the most appropriate strategy for implementation. The standardization of systems is seen as being particularly important as this would reduce the costs of equipment and help in the development of a European market which would in turn increase international competitiveness. The DRIVE programme links in with other European research under the EUREKA framework (such as Prometheus and Carminat), EURET, IMPACT, COST, ESPRIT and RACE.

In Japan, driver information systems have already been introduced into Tokyo with the primary aim of reducing traffic congestion. Each road user requires information about the destination, the actual traffic conditions on the selected route and alternative routes. An advanced traffic information system needs to supply this information, and two experimental systems have been introduced (Kashima, 1989). The bus location system transmits data via a roadside unit to a central computer which then instructs the bus driver at what speed to run and also transmits the arrival time of the bus to each stop. The system was introduced in 1983 and has resulted in demand increases of up to 30 per cent. The information on arrivals can also be obtained from the home and shops via local radio. Automatic vehicle monitoring system (AVMS) has been installed in taxis and trucks to monitor their position at any point in time. However, in Tokyo the density of beacons is not high and so the exact position of vehicles cannot be determined. The high costs of access to the system have resulted in only taxis participating fully. Installation has resulted in the number of vehicle trips increasing by 20 per cent (Kashima, 1989).

The most important developments have taken place under the Comprehensive Automobile Control System (CACS) set up by the Ministry of International Trade and Industry (MITI) with a 7.3 billion Yen budget (45 million Ecu: 1973–79) and a test site in south-west Tokyo. Time savings between guided and non-guided cars were about 11 per cent, and it was estimated that a system which covered all central Tokyo would give time savings valued at 80 billion Yen a year (500 million Ecu).

Two further systems were tested. Road automobile communication system (RACS) uses roadside communication units (beacons), in-vehicle units and a systems centre. This system does not supply continuous communication services, but intermittent communication over very small distances. But it can provide a service to the whole nation using one radio frequency and it was introduced on a freeway network in Tokyo in 1990. Advanced mobile traffic information and communication system (AMTICS) provides real-time traffic information collected by traffic counters and TV cameras, and these data are relayed to vehicles via a teleterminal system. The National Police Agency has installed 6200 traffic counters and 100 TV cameras to provide the information which is presented to drivers in the form of road maps. It requires each vehicle to be equipped with a standard digital road map database (Kawashima, 1990).

These systems are now being developed into full route guidance, but several unresolved questions still remain. It is unclear whether any system should be publicly or privately operated, who should pay for the installation of road side equipment, and the social acceptability of the technology (Tanaka, 1991).

In the USA, it is estimated that incident-induced congestion causes over half the total delays on urban freeways, and that this figure is likely to increase to 70 per cent by 2005 (Khattak *et al.*, 1991). The Intelligent Vehicle Highway System (IVHS) intends to use real-time information to guide drivers through the road network, and three linked programmes have been proposed (Chen, 1990).

- Advanced traffic management systems (ATMS) would include toll billing, incident detection and adaptive signal controls.

- Advanced driver information systems (ADIS) would cover vehicle location, vehicle navigation, information, route guidance and collision warning.

- Automobile vehicle control systems (AVCS) would be longer term and cover collision avoidance, speed/headway keeping, automatic highways and automatic guidance.

Chen (1990) concludes that the USA is behind Europe and Japan in both research and applications, and that progress in the USA is dependent upon the US Congress continuing to allocate a substantial budget for research.

Hepworth and Ducatel (1992) have neatly shown how the different groups of RTI technologies interrelate (figure 7.2). Traffic management systems are at the centre of road transport innovation, but even with all the available technology there is no solution to congestion, even if electronic road pricing were to be extensively introduced in Europe. There are also unresolved questions of responsibilities and priorities as they relate to the control of passenger information systems, the form of regulation and accountability. The availability of information and the control of that information are two crucial determinants of power. It is likely that 'infotactics' will be used to maintain that power (Toffler, 1991). The opportunity exists for greater control over cars and lorries through the use of RTI. On-board systems allow for the monitoring of vehicle performance and feedback to the driver. Traffic communication systems provide the link between the vehicle and the external traffic system. Finally, there are the driver support systems which give route planning and

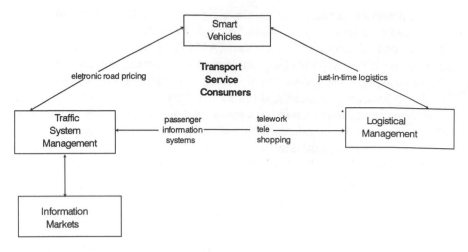

FIGURE 7.2 The Road Transport Informatics Family. (*Source*: Hepworth and Ducatel (1992))

guidance information to the driver. All three aspects form the components of the 'smart vehicle' (Hepworth and Ducatel, 1992).

### THE REALITY

To many people the opportunities offered by RTI must seem very attractive, but this technologically-led top-down approach to the problems of traffic congestion and network inefficiencies must also realize its limitations. In isolation, RTI will not solve these problems, and at best it may help to alleviate some of the particular instances of congestion, if combined with other transport and land-use policies. Bernard (1981) has identified three success factors in determining a technology's future market:

- technological superiority of the new substituting technology over the old;
- lifecycle cost advantage relative to the old technology;
- size of the potential market assuming 100 per cent substitution.

The reality is less clear than the conditions set out above. The RTI technology is likely to exist in parallel with the old technology, its costs (at least in the short term) are likely to be considerably higher than existing technology, and market penetration will be limited (at least initially). Given these uncertainties, it is unclear whether RTI will actually have any impact, particularly where consumer choice is involved. Its greatest potential would be in systems about which the consumer does not know (for example car engine management systems) and in business activities where there might be clear competitive advantages in using the technology. In both of these applications the car industry and the private sector can be expected to take the lead. There may be a role for information services to develop as an industry in their own right. In 20 years time over 90 per cent of the 190 million vehicles in Europe will have processors, communication and interface devices installed as standard. The EC is also concerned over the power of the USA which accounts for over two-thirds of the world's electronic databases. It wishes to develop a common EC industry to increase the existing market share of the UK, France and Germany which between them account for just 13 per cent of the world total (Vogel and Rowlands, 1990).

The EC DRIVE programme has defined seven areas for applications (table 7.6) which can also be related to figure 7.2 where some of the linkages are presented. In the management sectors there is considerable potential for RTI applications. This covers Integrated Urban Management, Integrated Inter-Urban Management, Public Transport Management and Freight and Fleet Management (Areas 3, 4, 7 and 6 in table 7.6). In each case the decision to use RTI will be commercially driven whether the responsibility lies with a private enterprise or a public authority. Even with reference to Demand Management (Area 1 in table 7.6), it is fairly clear that the public authority or the toll authority will use RTI. But it is here that such issues as the political accept-ability of the implications of using RTI (for example road pricing) become important. In the remaining two areas of applications (Traffic and Travel Information and Driver Assistance and Cooperation) most uncertainty re-mains, as questions such as user response, the costs of the technology to the individual, and public acceptability arise. It is also unclear as to who should pay for the infrastructure and who should control access to the information.

It may only be at the European level that such changes and decisions can be implemented, and this is reflected in DRIVE's aim of creating a pan-Euro-pean network of 'Information Centres' organized around distributed databases at the national, regional and local levels (CEC, 1991). These centres form a key element in the Integrated Road Transport Environment (IRTE). It seems fundamental that the administration and control of the data should be in the public sector and this means that the funding will also come from taxation, with a possibility of charging for the information. Public sector control would allow the information to be comprehensive and it would allow equal access to the database. Private sector control would create difficulties in data acquisition, and could result in inequalities in data access unless there was a powerful regulatory control agency. New organizational structures are required in road transport to ensure equal access to information by all competitors, whether development takes place in the private or the public sector, or a combination of both.

One major unresolved issue is the nature of user response to RTI in road transport. As Bernard (1981) has stated, assumptions are often made about the level of market penetration and the substitution effects of one technology for another. With RTI, the diffusion process is likely to be slow in terms of the awareness of the technology, in terms of the turnover rates of the car stock (typically 8–10 years), and in terms of the rapid obsolescence and the high initial costs of new technology.

Much discussion has focused on the particular problem of the impacts of telecommunications on travel, whether it results in the generation of more trips, the substitution of travel by communications, or a modification of travel patterns. The use of telecommunications to substitute for the journey to work has been seen as a strategy for reducing travel demand (Nilles, 1988). The most recent evidence from California (Pendyala *et al.*, 1991) demonstrates that the use of telecommunications does substantially reduce travel. On telecommuting days, those working from home make almost no commuting trips, with a 60 per cent reduction in peak period trip making, an 80 per cent reduction in vehicle miles travelled, and a 40 per cent reduction in freeway use. More important though was the tendency for telecommuters to chose non-work destinations closer to home and their overall 'action space' contracted after the introduction of telecommuting for all days of the week (not just those used for telecommuting). Non-work trips continued to be made at the same time as previously and to the same destinations. It seems that from the limited empirical data in California, collected at two points in time (1988 and 1989), lifestyles and activity patterns can be modified as a result of a major change in work patterns, and that these changes result in a net decrease in travel and the use of resources. If the results of the introduction of technology can lead more generally to changes in activity patterns, then the implications are considerable for congestion, energy consumption, the location of facilities and transport planning.

RTI in transport is unlikely to appeal to all drivers equally, and ironically some of the main advantages result from all drivers not receiving the same information. For example, the competitive advantage of route guidance systems is to allow some users privileged information so that their journey times can be reduced. If all drivers had the same information the alternative route being suggested would be less attractive. The marketing of RTI should therefore be aimed at increasing consumer awareness of the opportunities offered by different RTI systems and in identifying potential market segments. Access to technology will lead to greater inequalities in society as not all people will have the knowledge or the ability to use it. Information becomes crucial as does the presentation of the technology to the potential user. Access to these networks is expensive at the individual level and at the company level. Toffler (1991) calls this information divide 'as deep as the Grand Canyon' (p. 366).

The conceptualization of responses to RTI is based in traditional transport analysis. It is hypothesized that the individual will make more or fewer trips, change modes, change destinations, change routes, consolidate trips or re-schedule them. These behavioural responses are typical and can be incorporated in conventional demand analysis. However, there are several methodological problems. The range of responses is large and the initial scale of any RTI application is likely to be small, and so measurement of change will be difficult. It also seems likely that many changes will not be measurable in the terms outlined here as the driver may choose to ignore the advice given in the RTI application, or the modification of behaviour and consumer satisfaction may be marginal, or two people when presented with the same information may react differently.

With respect to at least two areas of RTI application (Traffic and Travel Information and Driver Assistance and Cooperation), effective implementation will only take place above particular thresholds of acceptance. Imaginative means to increase market penetration and measurement will be required:

- EC standards for all RTI applications will have to be established.

- The possibility of free installation of RTI in vehicles, at public transport termini, and in homes may be considered together with leasing the equipment.

- Publicity and marketing programmes will be required to raise public awareness and acceptance of the need for RTI.

- Careful analysis is required of the exact information and guidance specifications needed by the users.

- An understanding of the range of reactions from the car driver and passenger to route guidance advice or instructions will have to be researched. Many of these reactions cannot be covered by conventional demand analysis and new qualitative methods will be required.

Behind all these concerns is the role and the symbolism of the car which should not be underestimated. Much of the attraction of the car has been its ability to give the user freedom and the opportunity to make his or her own decisions. Many RTI applications in driver assistance and cooperation can reduce that independence. This may result in rejection and the user making the deliberate decision to ignore the advice given, hence eliminating many of the benefits from RTI.

The likely impact of RTI on transport congestion will be limited, unless the limited findings in California can be reproduced on a much greater scale. The most effective forms of RTI will be those which improve traffic management in the system (including public transport management). These passive systems require no action from the user of the system, but increase the reliability and efficiency of the system itself. The second area of significant potential is in logistics and in the organization of production and distribution systems, principally in the road freight sector. Here, competitive principles can be applied to ensure the operational efficiency of companies is maintained, and RTI may be a crucial part of that competitive advantage.

## Infrastructure

The success of cities and regions has always been based on the quality of their infrastructure. This in turn requires a commitment over a long period of time to continued new investment and replacement of existing stock. Infrastructure is the durable capital of the city and a country, and its location is fixed. In the transport sector it includes roads, railways, communications (for example air traffic control) and terminals. Often the services which are obtained from the infrastructure have a spatial dimension (for example the distribution of the rail network), with the benefit from that service declining as distance from the supply point increases (for example stations). Their key characteristics are that many people benefit from a single infrastructure, that they can be used over and over again, that the infrastructure remains when people and businesses move in and out of an area, and that it provides the means for integration and coordination of activities over time and space. The infrastructure forms the arteries of cities and nations, and the communications systems are the nerves. The health and prosperity of cities and nations depends on the quality of these networks.

A Europe-wide view of the transport infrastructure is necessary, and one which argues the case for massive increases in investment required to upgrade that infrastructure. This would include new methods for financing the infrastructure, new institutional and organizational structures, and a vision of a Europe-wide strategy of transport and planning. Much recent discussion has focused on the inadequacy of the existing infrastructure (for example Group Transport 2000 Plus, 1990). The costs of constructing new infrastructure and replacing existing infrastructure are considerable, and the massive investments in the 1950s and 1960s have been followed by stagnation and decline.

### THE PROBLEM

Since 1975, investment on inland transport has fallen in West Europe by 20 per cent in real terms and it has halved as a proportion of Gross Domestic Product (GDP) to 0.8 per cent. This reduction in infrastructure investment reflected general reductions in public expenditure, the world recession in the 1970s resulting from high oil prices, and the generally lower levels of increase in transport demand. Non-investment in transport infrastructure takes time to show an effect, and given the short-term time horizons of politicians, any delay in commitments to expensive projects meant savings in public budgets and lower taxes. Investment decisions were delayed, particularly expensive new links between countries and those which involved tunnelling.

With the economic upturn in the 1980s, there was unprecedented growth in transport demand, but it also became apparent that growth had continued throughout the 1970s as well. Investment had not been reduced because of reductions in demand for mobility, but for other macro-economic reasons such as pressures on public budgets, high interest rates, and industrial recession.

The 12 EC countries are second only to the USA in wealth creation (1986) and over the 20-year period (1970–1990) annual economic growth in the EC was 2.6 per cent per annum in real terms (Bendixson, 1989).

| European Community | 12 Countries | $3463 Bn | 2630 Bn Ecu |
|---|---|---|---|
| Western Europe | 18 Countries | $4163 Bn | 3160 Bn Ecu |
| East + West Europe | 25 Countries | $4739 Bn | 3600 Bn Ecu |
| United States | | $4185 Bn | 3180 Bn Ecu |
| Japan | | $1988 Bn | 1510 Bn Ecu |

Over a similar period, growth in passenger travel was 3.1 per cent per annum and growth in freight traffic was 2.3 per cent per annum, with the figures for growth in road traffic higher than those for rail. Since 1980, air travel in the EC has been growing by 6.1 per cent per annum. All forms of transport in the

TABLE 7.7. Evolution of transport in the Community, 1970–1990.

| | Transport in Billion tonne-km or Billion Passenger-km | | Market Share in tonne-km or Passenger-km | |
|---|---|---|---|---|
| | *1970* | *1990* | *1970* | *1990* |
| 1. INLAND | | | | |
| Freight | | | | |
| *Road | 377 | 797 | 55.0 | 73.9 |
| *Inland Waterways | 101 | 105 | 14.7 | 9.7 |
| *Rail | 207 | 176 | 30.3 | 16.3 |
| TOTAL | 685 | 1078 | 100 | 100 |
| Passengers | | | | |
| *Road (car/bus) | 1604 | 3089 | 87.8 | 87.9 |
| *Rail | 182 | 231 | 10.0 | 6.6 |
| TOTAL | 1786 | 3320 | 100 | 100 |
| 2. AIR | | | | |
| Freight | – | 196 | – | – |
| Passengers | 41 | – | 2.2 | 5.6 |
| 3. MARITIME | | | | |
| Freight | 85 | 100 | – | – |
| Passengers | – | – | – | – |

*Note:* Air figures estimates of traffic departing and landing inside the EC only.

*Source:* Based on Commission for the European Communities (1992*b*).

TABLE 7.8. *Evolution of road vehicles and road traffic in the Community, 1970–1987.*

|  | 1970 | 1987 | Yearly Growth 1970–1987 in % |
|---|---|---|---|
| Passenger cars in use (thousands) | 57459 | 116947 | 4.3 |
| Goods vehicles in use (thousands) | 7419 | 12881 | 3.3 |
| Vehicle-km passenger cars (billions) | 760.5 | 1399 | 3.7 |
| Vehicle-km goods vehicles (billions) | 157.6 | 275.4 | 3.3 |

*Source:* Commission of the European Communities (1992*b*).

EC have continued to grow (table 7.7), with the exception of rail freight. Most pronounced increases are apparent for air, road freight and road passenger travel (table 7.8).

That growth is expected to continue:

* Road and rail freight haulage will increase by 42 per cent and 33 per cent respectively (from a total of 805 billion tonne-km to a total of 1139 billion tonne-km in 2010).

* Car ownership will increase (to 2010) by 45 per cent (1985: 115 million cars; 2010: 167 million cars) giving EC averages of 503 cars per 1000 inhabitants (the 1990 level was 381 cars per 1000 inhabitants).

* Car travel will rise by 25 per cent from 1727 billion kilometres (1990) to 2166 billion kilometres (2010), and air travel will rise by 74 per cent.

The bare facts itemized here are only part of the problem:

* Demographic changes are taking place in Europe with the restructuring of households, more part time and female participation in the labour force, more working from home. If the motorization effect, increased levels of affluence and the grey revolution are added, there will be a substantial increase in levels of mobility (see pages 160–172).

* Technological changes will again seek to make travel easier, either through management and control systems applied at the network level, or through guidance and information systems purchased by individual drivers.

* Industry in Europe is being restructured with the switch from basic manufacturing industry to a service economy based on high-technology industry. The service sector accounted for 58 per cent of the EC GDP in 1985, and by the year 2000 this will have increased to 66 per cent. Most of the growth will be centralized at the core of the EC, but other factors suggest that the distribution of the industry will be more complex. There may be a migration of industry to Southern Europe (the sunbelt), where the quality of life is good and the costs of labour are cheaper.

- New manufacturing processes allow flexible location of industry and distribution centres, and the new telecommunications infrastructure also permits dispersion to accessible locations. However, underlying all these factors is the requirement for a high quality infrastructure which can accommodate the scale of changes itemized here.

The Group Transport 2000 Plus (1990) summarized the situation as follows:

> European transport faces a serious impending crisis. All the indicators point to this occurring when the Single Market is operational and at a time when there will be a massive increase in the movement of goods and services between the European Community and Eastern Europe. In all likelihood the crisis will paralyse the system, and slow down economic progress, provoke serious social tension, increase damage to the environment and destroy the balance in the central and peripheral regions of the continent. The process of building a unified Europe will be set back severely. Airports, rail systems, roads and urban centres have all faced a traffic growth at a rate far outstripping the increase in infrastructural capacity. (p. 45)

The cycle of production and consumption within the transport sector has been disrupted. Growth in demand (consumption) is often short term and relates closely to GDP and the levels of individual affluence. The trends over the last 20 years in the demand for travel have reflected these factors with the underlying growth being enhanced or diminished according to the macro-economic circumstances. The investment (production) cycle is long term and requires a commitment over a period of time. The lead time required from project generation, the construction time, the continuous process of maintenance and the ultimate replacement of the transport infrastructure requires this continuity of funding. At any stage in the process the continuity can be broken with decisions on investment being delayed. The short term costs are small, but each delayed decision adds to the 'impending crisis'. One possible scenario would be that all infrastructure reaches the end of its useful life at the same time, as decisions on new investments and maintenance continue to be delayed.

## THE CHOICES

Transport was identified within the Treaty of Rome (1957) as one of the areas for development of a common policy. Since that time progress has been slow with the Single European Act (1986) requiring the removal of physical barriers (at frontiers), reductions in technical barriers, and the harmonization of fiscal barriers. In 1985 the European Parliament asked the European Court of Justice to recognize officially the lack of a European Transport Policy and that this failure was due to the inefficiency of the European Council of Ministers (Commission of the European Communities, 1985). In 1986, the Commission put forward proposals for a medium-term plan on transport infrastructure (European Parliament, 1991). It described the principal deficiencies of the

European transport network, the ways in which the Community could take action to resolve them, the ways in which the Community could declare an interest so that Community action would be possible, and it identified the needs for overall financial investments in infrastructures. The Council of Ministers was reluctant to accept that proposal and in 1988 the Commission submitted a four-year plan extending to 1992, which coincided with the introduction of the Single Market. Again, there was resistance from the Council and the Commission presented more modest proposals which concentrated available resources on a limited number of projects regarded as the most important. This proposal was accepted (November, 1990).

Many international initiatives concerning transport have not come from the EC but from the industry itself. For example, the proposal for an international network of high speed trains has come from the Community of European Railways. The role of the EC has been of secondary importance and restricted to issuing Directives such as those on the environment, standards for road freight, cabotage, and reductions in customs' formalities. The Maastricht Agreement (1991) has expanded transport's role to include common rules on international transport and improvements in transport safety.

The EC budget for transport infrastructure investment is limited and any increase in that budget has been resisted by Ministers as there is a conflict with national interests and the notions of subsidiarity. Contributions have been made to specific projects, often under the European Regional Development Fund (ERDF) and European cohesion programmes. The ERDF allocated 43 per cent of its expenditure (1983–87) on infrastructure projects of Community interest, principally in those regions eligible for ERDF funding (Vickerman, 1991). Only 40 per cent of EC territory is eligible for ERDF financing, and transport infrastructure projects are recipients of substantial funding. In Greece 24 per cent of ERDF funding investment is on transport infrastructure, and the corresponding figures for Portugal, Spain, Italy and Ireland are 18 per cent, 47 per cent, 10 per cent and 39 per cent, respectively (European Parliament, 1991). The potential for private sector input has not been developed fully in either the road or rail markets. It is only in the air and telecommunications markets that the private sector has been fully involved, and it is here that most of the new investment has taken place (table 7.9). The possibilities for an enhanced role for the private sector either on its own or through joint ventures with the public sector have not yet been fully explored (Gerardin, 1989a).

The other main agency has been the European Investment Bank (EIB), which co-finances projects (up to a maximum of a 50 per cent contribution) designed to modernize Europe's economy. Between the date of its establishment (31 December 1982) and 1985, the EIB allocated just over 20 per cent of all its financing operations within the Community to transport. More recently (1986–1990), the EIB has further increased its support to over 15 billion Ecus (37 per cent of its total budget) for transport and telecommunications infrastructure and equipment (table 7.9). Assessment of projects is based on financial criteria related to the potential profitability of the project. The total allocation to transport and telecommunications has increased (Abbati, 1986):

| 1986 | 1945 Million Ecu |
| 1987 | 1661 Million Ecu |
| 1988 | 2980 Million Ecu |
| 1989 | 4001 Million Ecu |
| 1990 | 4518 Million Ecu |

To enhance European integration, the EIB has given priority to particular categories of projects (Abbati, 1986) which:

- help to develop regions in difficulties;

- achieve energy savings or other energy related investment so the EC's dependence on oil can be reduced;

- assist in European economic integration or towards the achievement of Community objectives like protection of the environment;

- modernize and promote sectors with high innovation potential including advanced technology.

EIB loans are generally repayable over 8–20 years, and loans can be backed by financial guarantee or by the assets represented in the project itself.

TABLE 7.9. EIB financing for transport and telecommunications.

| Country | Overland Transport | Air Transport | Shipping | Telecomms | Total |
|---------|---------|---------|---------|---------|---------|
| Belgium | - | 6.0 | – | - | 6.0 |
| Denmark | 476.1 | 241.2 | 5.6 | 188.5 | 991.4 |
| Germany | 252.3 | 30.1 | 1.2 | - | 283.5 |
| Greece | 208.0 | 7.4 | 0.9 | - | 216.3 |
| Spain | 594.7 | 652.1 | 40.3 | 1203.8 | 2490.8 |
| France | 2448.5 | 45.1 | 15.8 | 85.3 | 2594.6 |
| Ireland | 150.7 | 144.9 | - | 135.3 | 430.9 |
| Italy | 1294.1 | 414.4 | 302.0 | 2311.5 | 4322.0 |
| Luxembourg | - | - | 1.6 | - | 1.6 |
| Netherlands | - | 367.3 | - | - | 376.3 |
| Portugal | 639.0 | 53.6 | 57.7 | 176.7 | 927.0 |
| UK | 640.3 | 731.0 | 77.4 | 44.35 | 1892.2 |
| Article 18 | - | - | - | 660.8 | 660.8 |
| Total | 6703.7 | 2693.1 | 502.5 | 5205.4 | 15104 |

*Notes:* Telecomms includes telecommunications and telecommunications satellites.
Article 18 projects are located outside the EC (submarine cables, satellites, etc).
Figures are for 1986–1990 in million Ecu.

*Source:* European Investment Bank (1991).

However, across the EC there is still no infrastructure policy. The long-term ʝans provided by the EIB support individual projects submitted separately by public and private promoters in each EC country. Although it is recognized that international travel is only a small part of overall travel, some corridors are already at capacity, there are many 'missing links' in the European network, and demand will increase dramatically with the Single Market. Some strategy seems to be essential, particularly when transport policy is set against competition policy, regional development policy, and environmental policy within the EC. Congestion in the core area of the EC has resulted from:

- a high-density population and a high concentration of economic activities in those regions;

- north-south traffic crosses through this area and it is concentrated in a limited number of congested corridors;

- rapid and recent development of east-west flows related to the political and economic changes in central and eastern Europe.

The quality of the transport infrastructure is generally high and within the core area there are several international airports, a high density of roads and railway services. But the growth in traffic and congestion frequently disrupts transport systems and generates more bottlenecks (Gerardin and Viegas, 1992). Figure 7.3 illustrates the key transport problems in Europe and figure 7.4

FIGURE 7.3. Summary of key problems in the EC. (*Source*: Group Transport 2000 Plus (1990))

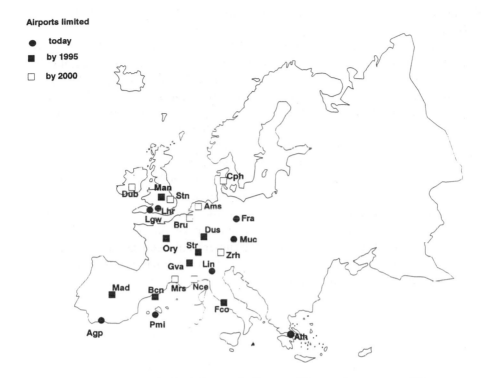

**Airports limited**

- ● today
- ■ by 1995
- □ by 2000

FIGURE 7.4. Frequency-limited airports in Europe. (*Source:* Association of European Airlines (1987))

presents the likely impact of the growth in demand for air travel over the next 10 years as more airports reach their frequency capacity. Half of the 46 major European airports will reach their capacity by 1996 and over two-thirds by 2000. Delayed flights have already increased from 13 per cent in 1986 to 23 per cent in 1989 (European Conference of Ministers of Transport, 1992). With respect to air traffic control, there is no overall system designed to coordinate all services. There are 22 systems operated out of 44 control centres with only limited coordination through Eurocontrol (Maggi *et al.*, 1992). Automated systems are available, but most contact is still verbal with communication being carried out through a multiplicity of languages.

Congestion is now endemic in both road and air travel. The consequences are well known and include reductions in effective capacity and mobility, increases in energy consumption and emissions of pollutants, and inefficient use of time. The social and economic costs of road congestion have been estimated at £15 billion a year in Great Britain (20 billion Ecu) and 1 billion Guilders for the Netherlands (Commission of the European Communities, 1992*b*). By the year 2010, with a 70 per cent increase in car use, the costs of congestion in the Netherlands could rise to 4 billion Guilder (1.8 billion Ecu).

The costs of airline congestion in Europe are now (1988) put at 1.4 billion Ecu (Commission of the European Communities, 1992*b*).

The Edinburgh Summit (December 1992) recognized the problem by setting up a new European Investment Fund to help fill the 'missing links' in European infrastructure and it also extended the lending facility of the European Investment Bank. It will now be easier for financial markets to back large infrastructure projects for trans-European networks, including transport, telecommunications and energy. The EIB will contribute 40 per cent of the fund's capital of 2 billion Ecu with the remainder coming from the Community (30 per cent) and the private and public sectors (30 per cent). It is estimated that the Fund could support investment projects worth up to 20 billion Ecus. The crucial assumption here is the expectation that the limited capital in the new Fund will generate this level of borrowing – the gearing ratio of 10 to 1 is very high. Nevertheless, this is an important step forward in EC thinking as it is the first time that a new financial instrument has been set up in this area. Previously, the Council had confined itself to laying down the objectives of an infrastructure policy and identifying the principal criteria for establishing whether that project is of Community interest.

Recently, the EC has argued for a Community strategy based on 'sustainable mobility' (Commission of the European Communities, 1992*a*). A new framework is proposed which sets out strict environmental standards for all modes of transport, for quality standards on pollution, for encouraging environment-friendly modes, and for the promotion of guidelines for infrastructure and the development of urban transport. These guidelines for the development and assessment of Community infrastructure projects would:

> discourage unnecessary transport demand and encourage where appropriate the development of alternatives to road transport, such as railway, inland waterways and combined transport. Guidelines for the conversion and upgrading of relinquished infrastructure, particularly for the purpose of 'soft' transport, would be implemented. (para 127)

The common strategy of 'sustainable mobility' should

> contain the impact of transport on the environment, while allowing transport to continue to fulfil its economic and social functions, particularly in the context of the Single Market, and thus ensure the long term development of transport in the Community. It should also contribute to social and economic cohesion in the Community and to the creation of new opportunities for the peripheral regions. (para 128)

In the EC White Paper on a Common Transport Policy (Commission of the European Communities, 1992*b*), one of the main themes has been trans-European networks. Incompatibilities between national transport systems have been highlighted, including inadequate interconnections, missing links and bottlenecks, and obstacles to inter operations, all of which lead to inefficiencies. The EC has had only a limited policy role, mainly through the Committee on Transport Infrastructure (set up in 1978), but with the principal financial contributions coming from the structural funds and instruments:

- ERDF credits for transport infrastructure 1975–1991 amounted to 16 billion Ecu;

- EIB loans for transport infrastructure 1982–1991 amounted to 14 billion Ecu;

- European Coal and Steel Community loans to TGV (*Train à Grand Vitesse*) track in France and Spain and to canals 1987–1991 amounted to 1.2 billion Ecu.

Now the EC proposes to establish and develop a

trans-European transport network, within a framework of a system of open and competitive markets, through the promotion of interconnections and inter-operability of national networks and access thereto. It must take particular account of the need to link island, landlocked and peripheral regions with the central regions of the Community (Commission of the European Communities, 1992*b*, para 140).

The goal is to improve the integration of the Community transport system and not the improvement of the transport infrastructure in general. It is likely that much of the funding will continue to be allocated to the geographically isolated regions.

On the crucial question of financing, the White Paper (Commission of the European Communities, 1992*b*) is pessimistic. The general level of investment in transport infrastructure has been stagnant at about 1 per cent of GDP. The volume of investment required for the period 1990–2010 is near 1,500 billion Ecu (1.5 per cent of GDP: Commission of the European Communities, 1992*b*, para 143). This level is far in excess of the resources available to the EC even if its mandate would permit such intervention. Its role is limited to financing feasibility studies, loan guarantees and interest rate subsidies. In addition, the EC may have a major dilemma. On the one hand it sees an under investment in transport infrastructure, but on the other hand it is arguing for sustainable mobility and protection of the environment. It could be argued that these two objectives are incompatible.

The EC policies on infrastructure now extend beyond the 12 Community members, and there is a specific provision for cooperation with third countries. The Prague Declaration adopted by the Pan European Transport Conference in 1991 emphasized the necessity of developing transport networks on a truly European scale and of integrating the greater European transport market. Measures have already been taken with the European Economic Area agreement and transit agreements with Switzerland and Austria. Trade between East and West Europe will increase movements in both directions by more than 50 per cent (1990–2000) which in turn will place considerable pressure on the links where little investment has taken place for the past 40 years (European Parliament, 1991).

The European Round Table of Industrialists (1988) has been one of the most vociferous in its criticism of the lack of a strategy on European infrastructure and among the most radical in its proposals. More generally, it seems that there is agreement over the problems and the options for infrastructure investment, and the next stage is to define the means by which priority can be allocated. Gerardin and Viegas (1992) suggest the following:

- Regional policy must be developed to balance the demand for and the supply of transport capacity so that future changes in those patterns can be accommodated. This means that adequate levels of funding have to be guaranteed so that the necessary infrastructure is ready at the time it is needed.

- European institutions have an important role to play in the coordination of regional policies, in the provision of investment, and in the dissemination of information between regions.

- Transport problems and solutions must be dealt with as a multimodal problem so that the best attributes of each mode can be used. These links are particularly important when Trans-European Transport Networks are being planned.

- Transparency must be ensured between the costs of providing the infrastructure and user charges. Modernization and construction of new networks in Europe require investment on a massive scale. User charges will increase but cannot assume the total burden.

- The private sector has a larger role to play and projects with joint funding will have to balance the internal profitability of the project with the positive or negative external effects.

- The long time scales involved in project development mean increased levels of risk, and so joint projects between the public and private sectors require a clear and well managed planning process.

- The regions represent the decentralized power of Europe and can respond to local requirements, and it is important to encourage collaboration between transport and regional development.

The proper role for transport in the process of economic integration in the EC has not been understood and short-term national interests have dominated. National transport policies may have reinforced existing barriers to integration (Vickerman, 1991).

# Transport Agenda 21

The three major issues facing European transport raised here do not give a complete picture, but they identify the main categories of problems. Some issues will be contextual as society is changing and this will have important transport implications (that is demographic change); some will be of a more speculative nature, but may cause fundamental changes in location decisions, activity patterns and lifestyles (that is technological change); and some will relate more directly to transport, but involve major commitments of resources at the national and international levels with both private and public sector participation (that is the new European infrastructure).

With respect to the transport issue, five important components of a policy on European infrastructure can be highlighted:

1. There needs to be a European strategy for transport and communications infrastructure investment which can be agreed by all 12 EC countries and followed over a period of time. In the past transport planning at both the national and international levels has been primarily concerned with the expansion of the physical capacity to meet the inexorable growth in demand. Even in time of recession it seems that demand continues to rise. Decisions have been taken on an annual basis as most transport infrastructure is funded from public expenditure, when continuity of both policy and funding over a period of time is required.

2. The Group Transport 2000 Plus (1990) suggested the imposition of a European Infrastructure Fund (agreed at the Edinburgh Summit in December, 1992) which would act as an autonomous source for all infrastructure investment with an EC dimension. Substantial sums can be raised from the motorist (estimated at 1 billion Ecu per year) through a carbon tax, and such a scheme could be expanded to other European countries on the same basis. Acceptance of the levy would only be gained if the monies raised were used for infrastructure investment and did not disappear into other uses. However, such a decision might contradict notions of subsidiarity, and it could be argued that most transport infrastructure investment should be the responsibility of national governments. Both the scale and the international nature of most transport infrastructure investment suggests that decisions need to be placed within an international strategy.

3. Often solutions are seen as relating to one mode of transport. Increasingly, multimodal options must be considered and here interchange points in the network which are most accessible will act as growth centres. This has already happened at international airports and at major nodes on the international road and rail networks (for example at Paris Charles de Gaulle airport and at Amsterdam Schiphol airport). Most recently, construction has started on La Cité de l'Europe, a large shopping complex in Calais aimed at users of the new

Channel Tunnel. The scheme has allocated 71,000 square metres to a range of retail and leisure activities at a location that is likely to see a massive growth in traffic as more than 18.5 million people are expected to cross the channel by coach and car after 1995, 13.5 million of them through the tunnel.

4.   Clear links must be established between a European infrastructure policy and the allocation of costs to users which fully reflect the external costs imposed. However, even here, there may be conflicts between this strategy and the broader requirements of providing infrastructure for social and regional development objectives. There needs to be a balance between these two fundamental EC policy objectives.

5.   Transport planning now needs to change its emphasis from its primary concern over the quality and quantity of the physical infrastructure to include other factors such as the organization, financing, technology, and environment dimensions. Methods are needed to explore the role that information technology can have in increasing capacity; the means by which a continuous flow of private and public funding, together with loans from the EIB, can be guaranteed over a period of time; the means by which the full costs of travel can be imposed on the user, either through road pricing, or through a levy (for example the proposed European Infrastructure Fund), or through a carbon tax; the appropriate balance between environmental concerns over transport and the considerable benefits that greater mobility and accessibility can bring to society; and the appropriate institutional structure for decision-making at the national and international levels. The establishment of the European Investment Fund is a key move in this direction.

The consequences of the growth in transport demand, the lack of investment in the infrastructure and the absence of a vision for Europe are now being paid. These factors, together with the changes in demographic structure, changes in industrial structure, and technological innovation have all resulted in the new mobility, high energy costs and environmental degradation. No real attempt has been made to apply consistent rules of competition in transport and the full environmental costs have not been paid by the polluter. Transport's role in integration of national and international economies and in assisting regional development has not been realized, and the potential for harmonization between modes has been neglected. The market has not been fully liberalized. Adequate, effective and efficient transport is a prerequisite guaranteeing the objectives of the European Community, but little progress seems to have been made.

# Chapter 8

# The Role
# of Transport Planning

Transport planning has now come full circle. Over the last 30 years its role has developed from analysis based on open-ended inductive experience to large-scale systematic analysis involving the systems approach to transport planning. It was only at the end of the 1970s that the most significant changes took place, with the initial broadening of concerns over the distributional effects of decisions, with the environmental concerns and the externalities of transport, and with the difficult decisions resulting from the restructuring of the economy. These policy concerns meant that transport analysis, at last, was seen as a central political issue and not one that could be left to the expert. In the 1980s, the planning framework was also dismantled as the principles of the free market economy were imposed, as transport enterprises were returned to the private sector, and as public expenditure was cut back.

The lessons from the recent past might suggest that there was no role for transport planning. This conclusion would be mistaken. As described in Chapters 6 and 7, there has been a renaissance in terms of the techniques and approaches used, and in the problems being addressed. Closer links are being established between land use and transport, new behavioural and dynamic techniques have been developed, the use of geographical information systems and graphical user interfaces is now becoming common, and a wide range of databases are being more extensively explored. The key problems of demographic change and industrial restructuring are being supplemented by demands for new infrastructure and the potential for the new technology in transport is now being comprehensively tested. These changes represent the European and national agendas. At the regional and local levels there are also key transport issues to be addressed, particularly where the market cannot work but a service needs to be provided. It is here that planning needs to operate as an alliance between state and the market. 'The market cannot produce optimal solutions for efficiency, nor socially desirable solutions in terms of equity' (Lichfield, 1992, p. 155). It is this alliance that is now discussed together with the levels at which and the priorities with which transport planning should operate.

## The Market-State Relationship

At present there is no national transport policy in the UK apart from the statement on road policy (Department of Transport, 1989*b*). The key relationships, such as those between road infrastructure and economic development, the narrow interpretation of environmental improvement, and the crucial links between land use and transport are not debated. According to the Transport Committee's Report on Roads for the Future (House of Commons, 1990) none of these is given prominence. At present the County Councils each produce a Structure Plan which deals with all policy aspects of planning including county roads, and the Department of Transport inform County Councils about substantive national road proposals. Part of the process involves public consultation with Examinations in Public being held. District Councils interpret the Structure Plans through Local Plans and Unitary Development Plans which fill in the details and allow large developments to take place even where road capacity is insufficient. The formal procedures are supplemented by Regional Planning Guidance and Planning Policy Guidance, and by Circulars which interpret the legislation. There is no detailed vision of transport, planning and land development potential or proposals for any location, city or county. The system is effectively a two-tier structure with the government acting directly with the District Councils and Statutory Undertakings (for example British Rail) on most transport and planning issues.

Decisions here need to be made at the strategic level, but it is here that the planning system is weakest. The new system of Regional Planning Guidance (RPG) offers the opportunity for integration within each region, but it does not integrate between the Departments of the Environment and Transport. It is solely the responsibility of the Department of the Environment (ACC, 1991). Although Development Plans have been strengthened by the recent Planning and Compensation Act, County Structure Plans are less important than they used to be, and Counties no longer have the coordination role for public transport. Too much responsibility for local policy on the land-use transport interactions is given to the District Plans which are implementation plans, not strategic documents. A recent study by the County Surveyors' Society (para 9.18, quoted in ACC, 1991) showed that in 1989 over 10,000 applications which had important implications for highways were dealt with by District Planning Authorities.

Some form of planning is essential, not based on conflicts between central and local government or on the dichotomy between the public and private sectors. It should be based on an alliance. Various forms of partnership must be identified and the case for such an alliance is clear.

1. There is a need for *intervention in the market* with new forms of regulation. Even the private sector favours a strategic framework within which to operate as stability in political objectives reduces uncertainty and risk, and these two factors in turn allow greater scope for private sector action. Such a strategic view does not conflict with the free market as it would still permit businesses to pursue their own profit-related objectives. The publicly-planned strategic framework would provide the longer-term horizon.

2. *Environmental issues* must now be seen as integral to all transport policy decisions and investment proposals. Decisions must be based on a balanced view between value for money and the externalities created. If a serious move towards sustainable development is to be achieved, then economic evaluation has to be moderated by political accountability as there is a considerable cost to becoming sustainable. In the Netherlands the current figure for the costs of their environmental policy is put at between 2 and 3.5 per cent of Gross Domestic Product. This level needs to be doubled to achieve the strategy set out (see pages 116–119). All forms of transport could be given environmental targets with green taxes or green grants being used to ensure all users are aware of the costs and benefits of using particular modes. Environmental audits could be carried out on firms and schemes as they are implemented, to achieve the minimization of total resources used, the maximization of recycling, and minimization of damage to the environment. Such an audit could take place on proposals to close facilities as well as decisions on opening new facilities.

3. Underlying much of the debate over the need for radical change is the *political acceptability* of any set of proposals. To make any real change in people's travel patterns and to reduce the use of the car, a fundamental switch in attitudes is required. People do not seem to be prepared to make a change in their lifestyles, as any form of commitment falls short of real action. No matter how attractive public transport is, no matter how close facilities are located to the home, no matter how expensive petrol is, people will still use their cars. Policy levers such as pricing and control may only have a limited effect on car drivers' behaviour. To hold any expectations that reality is different is unrealistic until public attitudes and priorities change. There is evidence that this change has taken place to some extent in European cities (for example Engwicht, 1992), but in at least as many other situations such a change is becoming more difficult as more people acquire cars and as lifestyles through choice become more car dependent. Perhaps this task is too difficult and should not be attempted and the politicians should continue to muddle through rather than take a decisive lead.

4.  *Traffic degeneration* must form an integral part of any thinking. It is now accepted that as new road capacity leads to a reduction in travel costs, more travel will result, and this in turn will lead to changes in accessibility brought about by changes in land use and the location of activities. New traffic generated as a result of new capacity forms one of the most difficult parts of current road assessment. In most cases, it is ignored and assessment concentrates on the redistribution of existing traffic supplemented by the secular growth trends in traffic. The use of fixed travel matrices is a major limitation in current analysis methods. Equally important is the understanding of the conditions under which traffic is actually lost, either through restricting the use of the car (for example by traffic calming), or through land-use and development policies (for example by concentrating on city centre development), or through the use of technology (for example by telecommuting). As the quality of travel deteriorates people may reassess the necessity to make the journey.

Radical alternatives are required. Transport systems management was used in the 1970s to increase the capacity of the road transport network through low-cost schemes such as area traffic control, restrictions on parking, and extensive one-way systems. This was followed by demand management in the 1980s to promote car sharing, new public transport systems, parking controls and pricing, and extensive pedestrianization and calming schemes. In the 1990s, the options are reduced as the technological problems have been solved but there is an institutional reluctance to implement a road pricing policy or to use planning policy to limit the growth in traffic at source.

The question of public acceptability seems not to have been tested in Britain. There is an awareness of the problems caused by congestion, on the economy through wasted time and resources, on the environment through levels of pollution, and on the health of individuals through increased stress and frustration. Yet no government or local authority seems to be prepared to tackle the problem, either in terms of primary legislation which would be required to introduce road pricing (for example a requirement for vehicles to be fitted with a metering unit and for the right to charge for the use of the road), or using the already considerable development control and planning gain powers available to local authorities. There is a fear of public reaction and of competitive disadvantage if only one city made such a decision.

In other countries the problem of public acceptability has been tested through a referendum. The city of Zurich, the state of California and, most recently, Amsterdam have all used this means to assess whether political action would be supported by the electorate. California has taken the lead in imposing strict environmental limitations on car emissions, in encouraging zero emission vehicles, in using land-use designations to reduce levels of trip generation, and in investment in public transport. Most emphasis has been placed on the journey to work problem and employers are seen as having a clear role to play in reducing trip rates. It is ironic that many of the radical proposals are being tested in the USA where the car culture is most dominant.

In Europe there is greater acceptance of restraint on the use of the car and there are strong planning controls. However, the concept of the compact city, advocated recently by the European Community (CEC, 1990), is continuously being weakened with peripheral developments, business and science parks, new towns, and the growth in international travel. The traditional radial patterns of movement are disappearing as cities change and as commuting ceases to dominate. Growth has taken place in circumferential movements, in new trip patterns for social and recreational activities, and in longer trip distances, and the net result has been the development of an enormously complex set of travel patterns.

## The Question of Scale

Planning must operate at all scales and provide the framework within which the market can operate. It must modify the market where there is a necessity to balance efficiency against other objectives. This role would be complementary to the political objectives set at the national and international levels which would ensure that fair competition takes place, that quality standards are observed, that regulations are set, and that barriers to movement are eliminated. The responsibilities of planners extend beyond the narrow interpretation of planning in terms of physical and economic planning. These responsibilities would include issues such as planning for institutions and finance, technology, energy, the environment and planning for people. These themes should cut through the basic grouping of factors identified in table 8.1 and they have impacts at all levels.

Within this framework, this Chapter concentrates on these major issues at four levels where it is seen that transport planning has a major role:

- economic growth;

- promotion of economic development;

- cities for people;

- meeting social needs.

These issues relate to the middle column in table 8.1 and form the central focus of the means by which transport planning can be linked in with the changing demographic and industrial structure of society, the city as a place, and the distributional impacts of decisions. It is at these three levels that transport planning still has an important role in determining the quality of life into the next century.

TABLE 8.1. The basic physical and economic components of the planning system.

| Scale | Issues | Methods |
|-------|--------|---------|
| European Community | Economic growth | Macro-economic Models of the real and monetary economies |
| National | Regional and economic development | Forecasting models Geographical information systems |
| Regional | | Input-output models |
| Local | Accessible and liveable cities | Land-use and transport models |
| | Distribution | Behavioural models Expert systems |
| | Needs | Minimum levels of service |

*Note*: The issues and methods may operate at different scales but have been allocated the most relevent scale.

## ECONOMIC GROWTH

For the most part this book has focused on the problem of transport from the view of transport analysis. One of the main criticisms made throughout is that such a focus is unnecessarily narrow as travel is dependent on land use and urban structure and the characteristics of the population, as well as the transport network. To produce a truly holistic view of transport demand it is necessary to integrate transport models with the broader perspective of macro-economics and the property market. Such an approach would not, however, be targeted upon the full spectrum of levels of policy interest in travel demand. It would be confined to the international, national and strategic regional levels, both of analysis and of policy formation (table 8.1). It would not be so appropriate for local and urban scale analyses concerned with the details of network loading and mode choice, but should address the longer term questions of aggregate regional and sub-regional infrastructure requirements.

The argument for broader based approaches to strategic analysis is twofold:

1.   Travel demand is, in most cases, derived from other demands – from wishes or needs to participate in other social and economic activity. It is a paradox that, whilst this proposition is widely accepted, transport researchers have persisted in attempting to develop more or less comprehensive models which effectively detach the analysis of trip making from all but the simplest of socio-economic frameworks. Long-term travel demand changes should

be analysed as derivatives and should therefore be modelled by drawing selectively from the fields of economic, demographic, property market, and land-use analysis, rather than autonomously as at present.

2.   Large-scale comprehensive models have not proved conspicuously successful in the past. They have been found expensive to develop, calibrate and maintain (due to their extensive data requirements and unwieldy structures), and have often been seen to exhibit implicit properties which tie their operability to particular phases of the long-term cyclical pattern of economic development. Classic examples have included Lowry's (1964) *A Model of Metropolis* and Forrester's (1969) *Urban Dynamics*.

The primarily derived status of travel demand becomes the lynch pin of a much looser modelling framework – one in which the outputs of major non-transport forecasting projects are combined in such a way that medium-to long-term shifts in national, regional and sub-regional travel demand can be modelled as derivatives speedily and without massive data gathering and calibration exercises.

Two such approaches can be identified which would link transport analysis with models of the 'real' economy and the 'money' economy.

## Models of the 'Real' Economy.

These models are based on macro-economic forecasts of the economy, together with its main regional subdivisions. Two macro-economic standpoints are reflected:

1.   Three models (those of the City University Business School, Liverpool University Macroeconomic Research Unit and the London Business School Centre for Economic Forecasting) are essentially monetarist in orientation, emphasizing the financial sectors of the economy and the links between money supply, interest and exchange rates. They are typically based upon the assumption of national expectations and are all built using straightforward econometric procedures.

2.   The other two principal models (Cambridge Econometrics and the National Institute of Economic and Social Research) have somewhat longer pedigrees and reflect a regression rather than monetarist economic disposition. They emphasize 'real' rather than monetary variables (demand, employment and output). That of the NIESR is of a traditional econometric structure, but the Cambridge model is a latter-day derivative of Stone's multi-sector dynamic input-output model, much extended and embellished in recent years.

Of the two alternatives, it seems that the second, based in a Keynesian structure, has most potential for travel demand analysis:

- It allows longer term forecasts over 5 to 10 years whilst the monetarist models are short term with quarterly (sometimes monthly) forecasts over a 12-month period.

- A real economy model which offers detailed sectoral disaggregation (for example on employment and consumer expenditure) is close to a necessary condition for the production of regionally partitioned forecasts.

- A close match can be achieved between the disaggregate output measures above and those which are likely to influence demand, such as demographic factors. The regional balance between labour supply and demand – absolutely critical to the estimation of travel demand – can be established from publicly available sources.

- If the travel demand that is initiated by the regional economy is mediated through the pattern of land uses and the process of land-use change, then it becomes essential to translate economic forecasts into property market projections. Relative shifts in the commercial and residential property markets underpin the processes of land-use change, and in an increasingly unfettered way as the planning regime becomes ever more permissive.

*Models of the 'Money' Economy*

The case for relying heavily upon real economy models, stepped down to regional and property market levels, appears close to overwhelming. However, the partial position of the commercial and residential property markets in determining the land-use impacts upon travel demand does raise some further questions.

Regional and sub-regional property markets represent the outcomes, not just of user and occupier demand pressure, but also of independent supply side effects. Adjustments to rates of development are not just mechanical responses to shifts in rental values. The partial autonomy of the property markets is evidenced through their propensity, over the past 20 years, to move in cyclical patterns which have not matched those of the wider economy. At the very least there are significant time lags.

If it is the case that the supply side of the commercial property markets may get well out of step with underlying rates of demand, then millions of square feet of commercial space may get developed in a sub-region (for example London) well in advance of what is required. The result is likely to be a collapse in rents which might, in turn, be sufficient to shift the underlying pattern of regional economic relativities. And the travel demand consequences over the period of two or three years could be very considerable.

Thus there does appear to be a case for considering the alternative economic modelling perspective which emphasizes monetary variables and through these the independent movement in the supply side of the various equations. The reasoning is as follows:

1.   The agencies involved include most of the major financial institutions (insurance companies, pension funds etc.), the property development companies, the house building agencies, and the main commercial owner occupiers who are forced to monitor the returns on all their assets including their property holdings,

2.   The capital investment and disinvestment policies of each of these classes of actors in the property markets are ultimately governed by the risk-return ratios prevailing in the international stock markets. Fund managers are constantly comparing investment in direct property with returns on equities (UK and non-UK), gilt-edged stocks and other long dated bonds, and liquidity (that is return on the money-markets).

3.   Risk-return relationships are difficult to predict, but underlying their movements, the preliminary interest and currency exchange rates are critical. House builders and property companies, which together produce most of the country's new built stock, are predominantly debt financed, and so gearing ratios related to the costs of money are fundamental to their operations.

It seems, therefore, that a robust long-term travel demand modelling strategy would be based upon the use of at least two types of national economic model. The first would centralize the documentation of the real economy of demand, output and employment; it would be disaggregated to a regional and sub-regional scale; it would at this scale be linked with demographic models to balance labour demand and supply; and it would generate regionalized outputs appropriate to the calibration of property market models which are essentially demand driven.

The second type of model would emphasize monetary variables and the supply side of the national economy; it would link the analysis of interest and exchange rate movements directly into the process of investment asset allocation forecasting; and it would provide independent development pressure inputs to property market models.

COMMENT

Macro-economic change and the impact of technology pose different problems of analysis as the scale is national and the effects occur over a longer period of time. Consequently, it is argued that the short-term local-scale analysis methods normally used in land-use and transport analysis may not be appropriate. It is here that macro-economic models together with models of the property market would be useful. Changes in demand for transport arise as a result of changes in the overall economy, in its industrial structure, and in technology. Keynesian models of the economy allow regional factors, demographic changes, and employment effects to be incorporated in to the analysis for a time horizon of 5–10 years. Some of the approaches use an input-output model and can therefore explore the impact of industrial structure on location decisions.

On the other hand, monetarist models lay more emphasis on the money supply, interest rates and exchange rates, and thereby analyse shorter-term movements in some of the key supply side variables. Through investment asset allocation forecasting, flows of capital into the industrial and property markets can be estimated, and from this the levels of trip generation. Together, these two approaches could help in the estimation of total travel demand at the macro-level in the short and medium term, and transport analysis would be truly integrated with the economic and industrial structure at the international and national levels.

## THE PROMOTION OF ECONOMIC DEVELOPMENT

Planning has switched from the notion of directing growth through investment to that of promoting development. The infrastructure necessary for development is no longer provided at public expense so that developers can enjoy the benefits through higher rents. An essential part of the development process is that of negotiation between the public and private sectors through exploring the means by which cooperation can take place. Previously, planners had been engaged mainly in designating land uses, and in securing adequate public infrastructure, amenities and housing. The concern was over comprehensiveness and long-term strategic objectives together with the ideal that development could take place with the interests of all parties safeguarded. This ideal has long been discredited as the interests of particular groups (for example business and car drivers) are always promoted over those of others. Altshuler (1965) argued that planners only had limited knowledge and this made their understanding of the public interest problematical, whilst Harvey (1978) and Foglesong (1986) both concluded that the planning system was oriented towards capital accumulation and that diversion into equity considerations was limited to further legitimation of dominant economic and political interests.

The industrial restructuring of city centres and the decentralization of work and homes to greenfield sites have caused a radical reassessment of the role of planners as regulators. There is now a direct connection between the economic structure and planning legitimation (Fainstein, 1991). Ideology no longer obscures the planner's role and the difficulty now relates to the means by which benefits to private developers can be equated with public sector objectives. The arguments used by governments concerning the efficiency, accountability and productivity of private firms have not been countered by an effective alternative argument that activities traditionally carried out in the public sector should continue to be exclusively in the public domain. Partnership is essential, and the balance has switched to the private sector.

In the USA, planners have either been political agents, or have worked in public development corporations or in private consultancies. Formal procedures are not followed but a process of continuous interaction (Fainstein, 1991). Subsidies can be offered to industry willing to remain or expand in the city centre, tax policy can be manipulated, regulations can be loosened, infrastructure can be built, and deals negotiated to allow developers to trade public benefits for mitigation of regulations. The British system is more rigid with set procedures and a wide ranging documentation of strategies at the county and local levels. However, even here greater flexibility is being encouraged, particularly at the interface between the public and private sectors. The development orientation allows planning gain and in certain situations the planning system has been bypassed. Urban development corporations have been set up to facilitate rapid development, including land assembly, tax relief (in enterprise zones), and in principle speed up the necessary associated infrastructure construction. Perhaps British planners need to be given comparable powers to their counterparts in the USA to allow a much greater flexibility and powers of negotiation so that industrial and commercial development can be attracted to city locations. Rather than set out a series of long-term objectives together with the means to achieve those objectives, planners are now concerned with action planning – limited objectives, and an overriding concern with seeing things happen and seizing opportunities as they occur. Beauregard (1990) calls this process city building, where narrow goals are set and all the means available are used to permit progress. This is the city competitive based on market rationality. Notions of long termism, a rational comprehensive approach to planning, and the minimization of externalities become irrelevant when objectives are short term and involve negotiations with the private sector.

Cities are now looser agglomerations with the link between homes and workplaces being broken. Industry is also footloose with location being based on a wider range of factors including the means to reduce costs through relocation. The functions of cities are no longer based on manufacturing production but on the supply of services (for example retail, financial and educational). Globalization of national economies and the close interrelationships between financial, commercial and manufacturing markets have all led to fundamental economic restructuring based on the new generation of communications, database and information technologies.

COMMENT

It is within this dynamic continuum that planning must be placed. There are now many different means to achieve a particular objective. This can only be

realized through shorter-term market analysis, project and programme evaluation, and a negotiation process. Planners become trapped in the logic of the
bottom line, trading off regulations governing density, light and air (and in
the United States loan and tax subsidies) for private sector commitments to
stay or build or contribute to community betterment (Fainstein, 1991). Within
this process it is impossible to separate transport from the development
process as it has played an integral part in allowing deconcentration of cities.
The evolving land uses have increasingly become car based with little opportunity for shorter journeys where walk, bicycle and bus would be more
appropriate.

Newman and Kenworthy's (1989) global analysis of energy consumption
in different cities has concluded that car dependence can be reduced by
physical planning policies, particularly re-urbanization and a re-orientation
of transport priorities:

- increase urban density;

- strengthen the city centre;

- extend the proportion of the city that has inner area land uses;

- provide a good public transport option;

- restrain the provision of car infrastructure.

These physical planning policies need to be supported by economic policies
on land use, housing and the provision of infrastructure. Hanson (1992)
suggests an integration of physical planning, cost strategies, and coordination
of land markets, housing markets, and transport systems if cities are to
re-establish themselves as centres of economic development. At present, the
incentives are encouraging more sprawl and low-density development which
results in car dependence and a high use of resources. Cities have to be seen as
attractive places in which to live and work, and transport planning has an
instrumental role to play in achieving that objective.

CITIES FOR PEOPLE

The accessible city is disappearing. This imbalance is likely to continue with
further dependence on the car and higher levels of mobility, both of which
will lead to severe congestion as the capacity of the transport system fails to
respond. Problems are created for industry, the environment and for people
to lead fulfilling lives. Cities then become less attractive places in which to
live as decentralization takes people and jobs to peripheral car-accessible
locations.

To reverse these trends cities must become attractive with affordable housing
and high quality facilities and amenities. They must be seen as safe and pleasant

places to live. Transport plays a pivotal role in achieving such a city, and although the car may still be essential it must be seen as promoting cities rather than as one of the main reasons why cities have become hostile environments.

## The Car

The car is an inefficient user of road space and there is a range of options available to ensure a greater awareness of those limitations. These include road pricing, the use of smaller vehicles, and ensuring a greater time in use as it is when the vehicle is parked it is still using valuable space. Road pricing is now a real option but there are still questions over its implementation which require much more general public acceptance than is now apparent. There are also major questions which have not yet been resolved, including the distortions caused by company cars, boundary and land-use pressures, land values and rents, the impact on the economic viability of city centres, the geographical area of application, equity and distributional issues, the use of revenues raised from road pricing, and even the impacts on the car industry.

Smaller city cars (including electric vehicles) and more intensive use of taxis are a real possibility, and several manufacturers are exploring lean burn engines (for example Ford) and electric city cars (for example General Motors and Fiat). Provided that these vehicles are given appropriate tax advantages (for example no vehicle excise duty) and priority (for example privileged access to parking spaces), then there would be sufficient incentive for their purchase. Alternatively, the Californian requirements for 2 per cent zero emission vehicles by 1998 for companies producing more than 35,000 cars (increasing to 10 per cent by 2003) gives clear incentives to industry to develop cars more appropriate to city use. Companies not developing their own zero emission car can sell another company's car at a subsidized price and cover those losses through raising the price of petrol cars. Similarly, if a company sells more than the magic 2 per cent, credit can be claimed and those credits can then be sold on the open market to other companies to fulfill their quotas.

## Parking

Similarly, a market priced parking strategy which involved appropriate enforcement measures and controls over all forms of city parking would ensure minimum parking time and full costs were imposed on users. These parking charges would be equivalent to the office rent and unified business rate level for the particular area. This would mean a doubling of existing hourly parking rates in Central London (Banister, 1989). Effective parking policy must include enforcement and payment of Fixed Penalty Notices together with the identification of frequent offenders, whilst removal of vehicles has acted as a major deterrent to illegal parking (NEDC, 1991).

*The Bus*

The bus is the most efficient user of city road space, yet it has to share road space and this in turn reduces its efficiency. The urban road network could be designated for particular users. For example, in the city centre, 30 per cent of roads would be allocated to pedestrians and cyclists, a further 30 per cent to public transport and access only, and the remaining 40 per cent to general use. The proportions would not be uniform across the whole city but would relate to dominant land uses. A central shopping area might have a higher pedestrian allocation, whilst a peripheral office location might have only a limited pedestrian and public transport allocation, and a higher general allocation. The implementation of such a scheme could take place immediately, and the capacity of the bus system would be dramatically increased as scheduled operating speeds would be improved. There would be fewer cars in the city centre and the environmental benefits would also be significant. The real value of such a scheme would only materialize if a network of bus only routes were introduced and buses were completely separated from cars on 'Green Routes'. The 'Red Routes' in London are a partial step to such a position, as parking prohibitions are strictly enforced, allowing the free flow of traffic. But buses still have to share road space with other traffic. The 'Green Route' option is now being considered in London in the context of bus deregulation as a set of exclusive rights of way would probably encourage greater on-street competition between operators. At present, unreliability costs operators over £100 million per annum in London through extra costs and lost revenue. In addition, the loss in travel time is significant, amounting to £127 million per annum in London.

*Coordination between Modes and High Occupancy Vehicles*

As a part of this priority strategy, coordination between modes must be promoted, so that the necessity to bring a car into the city centre is reduced. Parkway stations and park-and-ride facilities can be located at peripheral sites or at suitable interchange points at accessible sites within the city. High occupancy vehicle lanes can also be introduced on main motorway routes into city centres to encourage car sharing. Firms can positively encourage car sharing. All companies in California with over 100 employees are now required to have a plan to achieve target levels of vehicle occupancy for the journey to work of 1.75 in the central business district and 1.3 in the suburbs. Fines are imposed for non-achievement and some 209 employers out of a total of 6,900 were in violation by August 1991 (Atkins, 1992). Transport solutions are multimodal and involve increasing the occupancy levels for each form of transport. Consequently, the financing of new investment and the subsidy of transport should be seen in the same way. This means a move away from single mode financing towards decisions which relate to transport policy improvements and reductions in congestion. Regulatory policies should be seen in the same way.

## Capital Investment

In addition to the management of the existing road space and the allocation of space to priority users, capital investment is still desperately needed to update the existing public transport infrastructure. Light Rail Systems (LRT) are now being constructed in several British cities with the first phase of the Manchester Metrolink opened in 1992. Larger scale investments are planned for London with the Cross Rail scheme and possible extensions to the Jubilee Line and the Chelsea-Hackney underground. In all of these proposals, the financing has proved difficult as the government has imposed strict limits on public expenditure and as private finance has not been made available in the required amounts. In most LRT schemes, the local authority has financed the infrastructure costs together with Section 56 Grants (Transport Act 1968) from central government, and the private sector will be responsible for running the services.

## Allocation of City Centre Road Space

Allocation of city centre road space to particular modes may seem a radical proposal, but in other respects this strategy is the only one available in the short term, and it has been adopted in part through pedestrianization and traffic calming. Traditionally, the case for both these options has been argued on safety and (more recently) environmental criteria. Congestion has not featured, and unless pedestrianization and traffic calming are carried out on an area wide basis, they may just divert traffic away from one area to another, redistributing traffic but not reducing congestion. The environmental benefits are not convincing as research has not been able to establish a link between traffic speeds and noise and pollution in the urban environment. Most evidence relates to residents' perceptions of improvements in environmental quality when traffic levels and speeds are reduced.

A recent survey of UK local authorities has found a rapidly increasing number of traffic calming schemes being implemented with a high degree of public acceptance (Environment and Transport Planning, 1991). The local authorities reported that in 80 per cent of cases residents were in favour. Car drivers accepted the measures in 66 per cent of cases, and businesses in 68 per cent. A 90 per cent response from all UK local authorities revealed that 221 schemes had now been implemented, with a further 164 in the planning stage. Almost all the schemes were in urban areas. Closer investigation found that most schemes used road humps or speed tables (63 per cent) and some means to reduce street widths (49 per cent), with landscaping and changed road surfaces being used in a quarter of schemes. Reductions in accidents were used to justify the investments, but most schemes were not part of an overall transport planning strategy for the urban area as a whole. They were promoted in isolation.

For transport planners, there seems to be a series of related problems in implementing all of these strategies at the city level, and the problems are well illustrated with respect to the issue of traffic calming. First, traffic calming only rarely seems to be introduced as part of a coherent transport strategy. Each individual application is dealt with separately on its own merits. There seems to be a clear demand from residents to 'calm' their local areas, but there is no established methodology to achieve a priority ranking for all possible applications. Local authority budgets are limited and there is a concern among both politicians and professionals that any implementation strategy should be fair. Where attempts have been made to assign priorities to different schemes, the political implications have resulted in their rejection. For example, in the Royal Borough of Kensington and Chelsea (London) priorities were allocated on the basis of traffic volumes, traffic speeds, accidents, aesthetic values, costs, and the size of the benefiting population. However, the results were not acceptable to the local politicians whose own local knowledge was at variance with the results of the method developed by the consultants (Frank Graham Consulting Engineers Ltd, 1991).

The second related problem is the impact of the traffic calming scheme on the urban area as a whole. Positive responses from those living in the traffic calmed area are more than outweighed by anger from those living in adjacent areas where traffic levels (and accidents) have increased. Traffic calming must be seen as an area wide traffic management strategy as traffic and accidents may just migrate to neighbouring locations. The debate on this problem is not clear-cut as some (for example Tolley, 1990) argue that traffic actually disappears and crucially does not appear elsewhere. The circumstances under which this interesting proposition might be supported need further investigation – 'building roads generates traffic, removing them degenerates traffic' (p. 115). Tolley's vision (1990) would be to use transport planning to reduce trip lengths, to encourage a switch to soft modes (walking and cycling) and public transport, and to 'domesticate' the private car.

*Integrating Land-Use and Transport Planning*

Perhaps the most important contribution to reductions in congestion is to acknowledge the close links between land use and transport, and to make a greater use of the planning system to control transport growth. The principal objective must be to maximize accessibility and minimize trip lengths. These twin policy objectives would guarantee the greatest levels of demand for public transport, cycling and walk modes. By ensuring the appropriate mixture of land uses, the availability of local facilities and employment, and good quality public transport, the greatest efficiency in levels of transport would be obtained and the levels of energy consumption per trip and per person would be minimized, even where trip generation rates are high. The environmental costs of transport would also be significantly reduced if the dependence on car travel was reduced and balanced communities were encouraged (Banister, 1992a).

In the Green Paper on the Urban Environment, the EC argues very strongly for a compact city as the solution to the problem of urban congestion, lower energy consumption and lower pollution levels, and for the improvement of the quality of life (CEC, 1990). It goes beyond the sole concern with environmental sustainability to cover the impacts on the natural environment and the quality of urban life, and it views the city as a resource which should be protected. The quality of life in European cities has deteriorated as a result of uncontrolled pressures on the environment. In addition, the spatial arrangement of urban areas has led to urban sprawl and the spatial separation of functions. These factors have undermined the compact city which in turn has reduced creativity and the value of urban living. The return to the functionally-mixed compact city is a solution, the only solution to these problems.

At present, when decisions are made on whether to approve new developments, traffic generation forms an important element but only on a project basis. Assessment of the traffic implications on the system as a whole for the complete range of individual projects does not take place. A comprehensive perspective would allow traffic impacts to be assessed on the plan and programme levels, not just on the individual project level. Transport is a key factor in influencing land-use and development patterns and it is also linked closely with economic growth; this applies to freight and passenger transport.

COMMENT

Clear links have now been drawn between transport congestion, the environment and the quality of life. All the evidence leads to one conclusion, namely that cities must develop as attractive, safe places to live, and they must provide affordable housing, employment opportunities for a range of skills, and a diversity of social and leisure opportunities. Policies can all operate in the same direction to improve the environment, increase efficiency and promote accessibility. Some of the technological and planning options outlined in the previous two sections present the case for a combination of economic and planning levers, placed in a changing technological context.

Traffic calming and pedestrianization are seen by some as the only feasible solution to the problems created by the car (Tolley, 1990). Limitations on the use of the car through attempts to slow it down result in enhanced safety and improved quality of life in urban areas. It is not possible or desirable to continue to build roads to accommodate the expected growth in the demand for road traffic. Increases in capacity encourage more travel and longer trips, and reductions in capacity may encourage less travel and shorter trips.

Promotion of coordinated public transport on exclusive rights of way (rail and bus routes) would provide fast, reliable and safe services and help change the traditional image of buses. One of the main benefits from deregulation has been the growth in minibus operations. In 1985–86 minibuses and midibuses accounted for 14 per cent of the British bus and coach stock. By 1990–91 this figure had increased to over 27 per cent (19,700 vehicles). These smaller vehicles fill the gap between the taxi and the large bus, they allow hail-and-ride operations, they increase the feeling of a personalized service, and they provide greater security to the traveller (Banister and Mackett, 1990). Flexible routeing, shared taxis and the greater use of technology through passenger information systems can all increase the attractiveness of public transport.

Congestion pricing and parking controls are 'auto equalizers' designed to remove built in biases and subsidies (Cervero and Hall, 1990), but in themselves will not resolve the problem of urban congestion. Many of the problems of congestion relate to non-recurring incidents such as accidents, lane closures and unpredictable events. In the USA, some 50 per cent of freeway congestion is caused by these types of incidents (Cervero and Hall, 1990). It is here that technology has a key role to play through driver information systems such as route guidance and variable message signs. Both of these systems give drivers real-time information on traffic conditions. Continuous monitoring of roads allows emergency vehicles to reach incidents in the shortest possible time, it can help minimize the length of any delay, and it can help slow down traffic and redirect if necessary. In the longer term, fully automated highways and small electric/hybrid city vehicles may replace the current generation of cars. These vehicles may be hired or leased rather than owned, and would complement other more traditional petrol/diesel cars which would be used out of the city or on non-automated highways.

Land-use policy can also play an important role in the integration of public transport and land use, in ensuring that jobs and housing opportunities are in balance, and in designing people-friendly environments. If these objectives can be achieved, journey lengths would be reduced and the potential for using environmentally friendly modes (bus, cycle and walk) increased. Higher density mixed use developments in cities will enhance accessibility. It is in the suburbs that the increasing level of congestion is greatest and the solutions seem less clear (Cervero, 1989). The suburb to suburb movements on circumferential routes are ideal for the car, but the road capacity is limited. One option is for new road investment, but there is likely to be considerable opposition to new routes within existing urban areas. Public transport alternatives are less attractive, as are the opportunities for road pricing. The main policy lever available is land-use controls which provide for mixed developments, which will reduce trip lengths, but even this option will not eliminate congestion. If the example of the USA is taken, congestion within the city is followed by similar levels of congestion in the suburbs and beyond.

## MEETING SOCIAL NEEDS

One of the major roles for planning is to ensure a fair distribution of resources according to clearly established political priorities and that this distribution allows all people to gain access to the facilities which they need. The problem with a car-based society is that not all people have access to a car, and even those that do have a car in the household may not be drivers or the car may not be available. By examining the travel characteristics of selected social groups it is possible to assess some of the variations.

In a comprehensive review of UK national transport policy for the period 1960–1988 it was found (Hay, 1992) that there has been relatively little reference to equity, fairness and justice (EFJ) by either of the two main political parties, and EFJ issues were more often brought forward by minority pressure groups within the party or by backbenchers in Parliament. Of a wide range of definitions relating to equity, fairness and justice, only two were widely used. Formal equality was raised with reference to who pays the costs of providing transport, and basic need was raised with reference to the provision of public transport. The former concept requires that like individuals should be treated in a like manner ('horizontal equity' in table 8.2), and the latter is the principle that people should be able to secure the minimum required to sustain life and health ('equalization of opportunity' in table 8.2). The use of EFJ concepts has widened since 1978 (Trinder *et al.*, 1991). In a similar study based on interviews (1989) carried out in the Shire Counties and Metropolitan Boroughs, similar results were found. Meeting basic needs and recognizing formal equality feature dominantly with both councillors and officers (Hay and Trinder, 1991).

It is clear that there is a wider recognition of EFJ issues among both politicians and planners, but there is less agreement on what should be done because neither party is clear how to tackle the problem in the transport sector. It is also clear that all substantive transport planning proposals will raise EFJ issues.

One of the key responsibilities of transport planners is to understand and resolve conflicts between interest groups, and the understanding of EFJ must be an integral part of those professional skills (Hay, 1992). It seems that a narrow view of EFJ has been taken at both the national and local levels. Here an attempt has been made to combine various elements of EFJ through the two principal definitions of formal equality and basic needs (table 8.2). These

TABLE 8.2. Equity and transport.

|  | Equalization of Outcomes | Equalization of Opportunity |
|---|---|---|
| HORIZONTAL EQUITY – FORMAL EQUALITY | Service distribution according to demand | Equality in service distribution |
|  | Service provided on commercial criteria | Service provided to all communities at a similar level |
|  | Market based with no subsidy | Standards or minimum levels of service |
| VERTICAL EQUITY – SUBSTANTIVE EQUITY | Service distribution according to need | Positive discrimination for particular disadvantaged groups |
|  | Travel concessions for elderly, young, disabled and other need groups | Special services |
|  | Subsidy provided to the user | Subsidy provided to the service |

It should be noted that this table extends the interesting work of Rosenbloom and Altshuler (1977) who identified three main concepts of equity as they apply to urban transport:

• Fee for service – to each according to his or her financial contribution. Service distribution according to demand.

• Equality in service distribution – to each an equal share of public expenditure or an equal level of public service, regardless of need or financial contribution.

• Service distribution according to need – to each a share of public expenditure or service based on need, as government has chosen to define it and taken steps to ameliorate it.

concepts are supplemented by those of substantive equality and procedural fairness, together with various notions of needs (for example need as demand and wider needs).

Equity can be conceptualized as an equalization concept which is concerned with the equalization of opportunity and outcome. Two further elements can be added to this basic dichotomy. Horizontal equity requires government to treat like persons alike in decisions concerning funding, the distribution of benefits and compensation. However, as Ellickson (1977) comments, likeness is a matter of degree and he suggests that horizontal equity can be breached where Michelman's test applies (1967): this requires a person to bear a loss and that this is not unfair 'if he should be able to perceive that a general policy of refusing compensation to people in his situation is likely to promote the welfare of people like him in the long run.' Vertical equity relates to the fairness in the distribution of wealth among different income groups, and as with equalization of outcome requires some assumptions on income distribution. In effect it is argued that governments plan public projects either to stimulate some measure of income redistribution or to improve the relative position of the disadvantaged through the equalization of opportunity. Table 8.2 summarizes the matrix of equity considerations. It should be noted that decisions in transport bear on each of our four components of equity, but in no coherent manner. With present policy, it seems that commercial criteria will prevail over other considerations, and this is reflected in its increasingly dominant position with the reductions in subsidy and competition in the provision of transport services.

The apparent simplicity and tidiness of the table conceals a much more complex reality. However, by presenting a flexible yet strong framework for the identification of the disadvantaged and an evaluation of alternative forms of transport in terms of the outcomes and opportunities available, a clear picture of the importance of needs based transport planning can be discussed.

One major role for politicians and transport planners at both the national and local levels is to set the agenda for transport based on concepts of equity. A theoretical framework such as that developed here provides the context, and service provision can be based on assessments of the needs for the various disadvantaged groups (Banister *et al.*, 1984).

In a recent study (Else and Trinder, 1991), local government councillors and officers were asked to state the four major transport issues in their particular authority. Nearly 75 per cent of the issues mentioned came within the infrastructure category (roads, railways, maintenance, traffic flow, parking, environment and road safety). A further 20 per cent concerned public transport (bus, rail and the disabled), with the remaining responses covering finance, planning and social policy. The most important single issue was investment in the road network within the jurisdiction of the various authorities, and in many cases members and officers mentioned particular projects. The second issue was traffic flow (that is congestion), which, although related to the first, was seen as a broader problem for which new road infrastructure was not the only solution. The other two issues which scored more than 10 per cent were bus services and road maintenance.

As might be expected, less importance was allocated to road construction and maintenance in Labour-controlled authorities, with more importance being attached to public transport and the broader issues. But in all cases it seems that the predominant view was transport as a separate sector to which policy priorities were allocated in isolation, not as part of an overall strategy for the city or county.

COMMENT

This isolation is one of the main obstructions limiting effective action at all levels of decision-making, as responsibilities are compartmentalized within and between ministries and local authorities. Many decisions have transport implications. One of the major transport trends in the recent past has been the increase in the use of the car and the growth in average trip lengths. A major part of this increase can be explained by the closure of local facilities and the trend for increasing specialization within facilities, so that the nearest opportunity is not the one that will actually be used. These changes are taking place in all sectors supported by strong arguments for economies of scale and scope with thresholds being set for minimum viability (for example in schools). These rationalizations have affected schools, shops, banks, magistrates courts, petrol stations, hospitals and many other forms of employment and services. In all cases there may be significant savings to the supplier of the service, but no account is taken of the increase in user costs. The resource costs of travelling to work or to a facility are borne by the user.

It is not difficult to calculate the increase in these costs. If, for example, a local hospital is closed, the additional costs of travelling to the next nearest equivalent facility may be calculated. It is not surprising that longer journeys and increased car dependence are two major outcomes of such rationalization of services. But these additional costs do not form part of the decision as they are not costs which accrue to the supplier of the service. Recently, there has been much debate over the possibility of road pricing and significantly raising the costs of travel to the individual. It is argued that the costs of travel are too low and must be raised, particularly in congested urban and motorway conditions. There has been no debate over the increased distances and costs incurred to the individual who is now forced to use the car to get to hospital, to take children to school, or to get to the nearest shops.

One important role for transport planners must be to take a more holistic view of what is happening at the national and local levels so that the full implications of decisions taken in any given sector can be interpreted. There must be a balance between the narrow market driven criteria for rationalization of services and the broader society based evaluation. In some cases the broader arguments may be sufficient to alter the decisions to close a local shop, school or hospital. Pervasive externalities suggest that the social costs may exceed the private benefits resulting from closure.

Unless some attention is paid to these broader planning issues, more travel will be made by car, trip lengths will continue to increase and more resources will be used. Notions of accessibility are replaced by notions of mobility. The possibility of travelling to local facilities by non-energy consumptive modes is reduced and the ideal of sustainable settlements becomes ever more remote. A balance has to be found. The only way to achieve this is with an overall vision for urban planning and transport. Group Transport 2000 Plus (1990) concluded that 'the necessary restrictions on car use and parking must look beyond prohibitions or dissuasive tolls and enable full integration of private cars in the transport system . . . People are 'trapped' into using cars' (p. 17). The real costs of increasing car dependence should be assessed as should the policies of reducing car dependence. At present neither is.

*Chapter 9*

# Conclusions

What then is the future for transport planning? Over the last 30 years it has proved to be robust and resilient to change, and transport planners have remained loyal to the systems approach with as much as possible of the analysis being formalized in quantitative models. However, two major factors have now led to a reassessment of the traditional approach. These factors are not based on the limitations and critique of established methodologies, but on the radically changed nature of the problems facing decision-makers.

The first relates to the changes in the political environment with the ascendancy of the market view over the social and welfare perspective. The Anglo-Saxon philosophy, which has been dominant in the United Kingdom for over 10 years, reduces the role for strategic planning and longer-term objectives, and adopts a commercial (or quasi commercial) approach to maximize the internal efficiency of the transport system. The market mechanisms are given priority, but it is still realized that there are potential market distortions (for example monopoly power and externalities) which must be corrected by government intervention. Demand must be met at the lowest cost. This philosophy is dominant in the United Kingdom, Denmark, Spain, Ireland, Portugal, Greece, Norway and Sweden.

The alternative approach is the Continental European philosophy where transport is seen within a wider social perspective and there is a *Droit de Transport*. Here, a longer-term view is taken of the role of transport in economic development and political integration, with clear investment priorities and a greater degree of master planning. Transport must form an integral part of the social infrastructure as it affects the distribution of population, employment, shopping, and social life, and it has a direct influence on health and the environment. This much higher level of government intervention is still the dominant philosophy in France, Germany, the Netherlands, Belgium, Luxembourg and Italy.

Over the next decade there is a likelihood of further convergence between these two views as more pressure is exerted on public budgets and as the politics of Europe continue to move towards the right. For example, more countries will introduce regulatory reforms within all forms of public transport through open access to the market, through privatization, and through tendering for services. There is a tendency for governments to withdraw from the provision, support and regulation of transport services.

The second major change is the increased awareness that transport planning must be seen as an integral part of a much wider process of decision-making. Too often in the past have transport solutions been seen as the only way to resolve transport problems. This was first evident in the justification for more road building in the belief that a shortage of capacity was the main reason for congestion. Increasingly, it has become apparent that more capacity, particularly in roads, generates more traffic and allows for the dispersal of activities, which in turn makes the provision of public transport services problematical. Transport planning must be seen as part of the land-use planning and development process which requires an integrated approach to analysis and a clear vision of the type of city and society in which we wish to live.

All decisions influence this pattern of development, and if we are serious about achieving targets for reductions in emissions, improved safety, and a better quality of life in cities, then these issues must be fully reflected in the decision-making process. In many cases, only limited value seems to be placed on environment, accidents and quality factors. Yet increasingly, these are the issues which people claim to be concerned about.

The transport planning process is still driven by the desire to reduce travel time and costs. The premise underlying this assumption is that people value quicker and cheaper travel. The irony here is that it is now being argued that travel is too cheap and that prices must be raised to account for other factors, particularly externalities. The importance of speed is also being questioned as drivers are being urged to reduce speed for safety and environmental reasons, and lower speed limits are now being introduced in residential areas and even on motorways. So the two main foundations of transport analysis (time and cost) are being questioned, yet it is not possible radically to change travel patterns, at least in the short term. Simply raising the price of one fairly inelastic commodity (such as transport), for whatever reasons, merely means that additional revenues are raised as taxation for the Treasury. It is only when price increases are matched with quality improvements in transport services (particularly public transport), together with policies which make it possible to reduce the number and length of journeys by car, will any real benefits be achieved. It is not surprising that the public are against increasing the costs of travel if the quality of service is poor and declining, and if there is no real way to alter their lifestyles as planning and development policies have encouraged more travel and greater reliance on the car. Even within the transport sector itself, the alternative modes should not been seen in opposition to each other and set up as mutually exclusive choices. Too little attention has been given to obtaining the 'best mix' of modes through multimodal journeys involving interchange at park-and-ride and other facilities.

## Ethics in Transport Planning

The debate over the role that transport planning should have in the future is likely to continue, as is the artificial separation of planning from transport, the links between economic growth and transport, and the question of whether the cost of transport is too cheap. Throughout this book, it has been stressed that transport planning has a key role to play in forming an alliance with market forces at all levels. The perspective taken has been that transport planning is an agent for promoting economic development and for making cities attractive places in which people want to live, particularly for those who are seen as being disadvantaged (Chapter 8). Underlying this perspective is a need for professionalism in transport planning and the ethics of honesty. Transport planners should not try to claim that their methods and approaches to analysis have all the answers, but they should be prepared to discuss the limitations of analysis, together with their inherent assumptions and data requirements.

Transport planning works at the interface between technical analysis and policy-making, but very little discussion has developed around the ethical position of planners. It is not sufficient to argue that transport planning is value free and neutral in its approach to analysis. This attitude, deeply rooted in the traditions of positivism, overlooks the value laden assumptions, the quality of the data, the simplification of the decision process, and the inadequacies of both analyst and decision maker.

> The purpose of planning tools is to provide systematic and neutral information to support decision-making, while the ethical content of planning is assumed to be in the definition of the problem and the weighing of information by decision makers. Public hearings and citizen comments on proposals supplement 'objective' analysis by providing personal and subjective views. (Wachs, 1985*b*, p. xv)

There are no methods which can truly be claimed to be value free and neutral in their approach. Presentation of analysis should acknowledge this situation, with the expert being honest about the limitations of the analysis. Transport planners are both technicians and politicians. On the one hand, transport planning is seen as apolitical with analysis and rationality dominant, whilst on the other hand transport planners see themselves as having a role in bringing about social change. Here, priority must be given to the principles and legitimation of change rather than confidence in the objectivity of the analysis. The code of standards in ethics means adherence to basic values and norms, with a greater openness at all levels of decision-making concerning the advantages and limitations of analysis. This process may weaken the position of the professionals, but by open debate they will receive the respect of the public. This is the choice to be made.

## Infrastructure and Finance

Undoubtedly, the most important issue in the next decade concerns the transport infrastructure in Europe, whether it is new infrastructure, an expansion of existing infrastructure, or the reconstruction of existing infrastructure. The basic political questions are the amount of infrastructure which should be provided – whether it should meet unconstrained demand or whether demand should be constrained – and how it should be financed. As described in Chapters 4 and 7, much debate has revolved around the role that the private sector should play. However, most transport projects are large scale, involve a high risk and have long payback periods. It is only where the private sector has a virtual monopoly position (for example in road bridges across estuaries) that real interest has been raised. Transport projects are notorious for their cost overruns, for technical deficiencies in construction and consequent high maintenance costs, and for optimism in their estimates of future demand – the Channel Tunnel project illustrates all of these problems.

In the past, the public sector has been the main contributor, but with the increasing costs of new transport projects, and with the desire of governments to reduce public deficits new sources of capital are required. The most obvious means to raise finance would be to charge tolls on existing motorways. It has been estimated (Keating, 1992) that tolls of 5 pence per mile (similar to current toll levels on French autoroutes) on UK motorways would raise about £2 billion per annum and its extension to other roads and to urban areas would raise £6 billion per annum. The key argument is that the tolls must be seen not as a tax but a charge which can be reinvested in the transport system.

The private sector is unlikely to invest in major projects unless the conditions are favourable and a reasonable rate of return can be obtained to satisfy shareholders. This means that low-risk projects would be most desirable so that the following conditions could be met:

- to maintain a virtual monopoly position;

- to meet high levels of existing demand and anticipated growth in that demand;

- to ensure a payback over the medium term (about 10 years);

- to have control over the levels of pricing for the new or upgraded infrastructure;

- to have some guarantee from government of the period over which revenues from tolls will accrue to the private sector.

In most cases, one or more of these conditions are not met and as a consequence the private sector has been reluctant about making any firm commitment.

potential lies. The public sector can anticipate the growth in demand which results from the growth in the economy, rising income levels, and from new developments. It can also assist in the land assembly and the public inquiry processes so that time from project inception to completion is minimized. In certain situations, it can also help with the costs of construction of the actual infrastructure, but in most cases the private sector will manage and run the facility, including setting the levels of tolls or fares. This partnership is already apparent in the Manchester Metrolink.

In addition, new private companies could be set up to both build and operate the infrastructure, and finance can be raised in a number of ways:

- through loans from the European Investment Bank and grants from the European Coal and Steel Community (which has already provided funds for the Channel Tunnel), and the European Community Regional Development Funds and the new Transport Infrastructure Fund;

- through transport bonds and other long-term investments (using pension funds) – these are extensively used in Japan;

- through tax incentives to the private sector by making their capital contributions tax deductible;

- through tolls and road pricing;

- through employment taxes (as in Paris and other French cities) or a tax on petrol (as in Germany and the USA).

At present there seems to be an impasse between the public and private sectors which must be bridged. Each has an important role to play in the construction, renewal and maintenance of the road, rail and air transport infrastructure. Over the next decade it is likely that new means of financing will be established, yet it should be noted that the public sector cannot withdraw completely. Much of the funding will still remain the responsibility of the public sector, but the role for transport planning must be to facilitate private sector and joint ventures through advice, predictions, land assembly and accelerated public inquiry procedures. The public sector still has a key role to play:

- where the market fails and intervention takes place for accessibility, distributional and equity reasons;

- where there are significant externalities involving the use of non-renewable resources, land acquisition, safety and environmental concerns;

- where transport interacts with other sectors, such as the generation effects of new developments and priorities given for regional or local development objectives;

- where transport has national and international implications, such as promoting London as a world city or maintaining high quality international air and rail links.

## Into the Next Century

Throughout this book, links have been drawn between transport and other sectors, in particular the planning and development processes. In addition, the focus has been on placing the UK experience in the wider European and worldwide context of the demographic and technological changes taking place in society. The late twentieth century is witnessing the second great economic transformation (Castells, 1990). The first was the switch from agrarian to an industrial mode of production, and this in turn led to the development of transport and the high mobility, car-based society. The second is the transition from an industrial mode of production to an information mode where the fundamental inputs are knowledge based.

Castells argues that the three major changes in the capitalist economy in the 1980s have been:

- an increasing surplus from production with a substantial growth in profits;

- a change in the role of the state away from political legitimation and social redistribution towards political domination and capital accumulation;

- accelerated internationalization.

The net result of these trends has been a sharp spatial division of labour, the decentralization of particular production functions, flexibility in location, and the generation of information innovation milieux. Space has become increasingly irrelevant as cities have spread and many routine information functions have centred on peripheral, low-cost and low-density sites. The innovation milieux are elite locations where high technology activities are centred, and these locations form the magnets for economic growth and dynamism (Hall, 1988). It is here that the global shifts in power have taken place, and it is through the control of technology that power will be maintained (Toffler, 1991).

What then are likely to be the key concerns for transport planners in the next century? Some issues have been highlighted in Chapter 8. The globalization of the capitalist economies will accelerate, and this will not just involve the countries of the West and Japan, but those of Central and Eastern Europe, together with other emerging economies on the Pacific rim on both the Asian and American sides. There will be a transition from centrally controlled and planned economies to market based economies. It is here that a clear understanding is required of both the links between economic growth and transport demand and between the operation of the monetary economy and transport market.

Technological change will fundamentally influence the location of economic development and the function and form of cities. Convergence of computing and communications will allow a user friendly interface for many transactions, for information, and for business and social activities. As Hall (1991) states, the potential for such a system is vast as

> it would allow almost infinite dispersion of informational activities across nations and continents, including the spread of home-work and telecommuting. (p. 19)

But equally, information and technology may cause divergence as knowledge to use the system is not universal, as costs of access to the high-quality broad-band communications systems will be expensive, and as control of these systems may reside in the hands of a few multinational companies. Power may be concentrated in existing centres of information exchange at accessible points on the network or in the few world cities where rapid innovation and high-level service competition can take place. Second-order cities and those in peripheral locations will continue to be centres of low demand and low innovation.

Similar developments can be seen in transport where key centres in Europe are likely to be located on the high speed rail network, particularly at interchanges between road and rail, and between road, rail and air. The international nodal points will be located where the global airlines have their hub operations, and where good quality transport links are available to the local and national centres of population and activity. Increasingly, these nodal points will also be centres of the international information networks – the logistics platforms.

As with the industrial revolution, the technological revolution is likely to promote concentration, but at an international rather than national level. Smaller national centres in the hierarchy will be important, but at the national and local levels. It is difficult to speculate on the exact locations of the new global and national city hierarchy. However, the demand for transport will increase in scale, in range, and in quality as this globalization takes place. The House of Lords Select Committee on Science and Technology (1987, para 6.32) summarized the problems facing the transport planner:

> Transport needs depend on how and where people live and work. But changes in how quickly or conveniently people can travel even over short distances affect where and how they work and live, and if people change how and where they work and live, different transport needs will arise. The question is, how far the interaction between these things can be analysed.

This is the challenge.

# References

Abbati, C. degli (1986) Transport and European Integration, Commission of the European Communities, European Perspectives, Brussels.

Adams, J. (1981) *Transport Planning: Vision and Practice*. London: Routledge and Kegan Paul.

Alexander, E. R. (1984) After rationality what? *Journal of the American Planning Association* **50**(1), pp. 62–69.

Altshuler, A. (1965) *The City Planning Process*. Ithaca, NY: Cornell University Press.

Ambrose, P. (1986) *Whatever Happened to Planning*. London. Methuen.

Andersen Consulting (1990) *The Impact of Environmental Issues on Business – A Guide for Senior Management*. London: Andersen Consulting.

Appleyard, D. (1981) *Liveable Streets*. Berkeley and Los Angeles: University of California Press.

Association of County Councils (ACC) (1991) Towards a Sustainable Transport Policy. Paper prepared by the Association of County Councils, London.

Association of European Airlines (1987) *Capacity of Aviation Systems in Europe: Scenario on Airport Congestion*. Brussels: AEA.

Atkins, S. (1987) The crisis for transportation planning modelling. *Transport Reviews*, **7**(4), pp. 307–325.

Atkins, S. (1990) Personal security as a transport issue: A state of the art review. *Transport Reviews*, **10**(2), pp. 111–124.

Atkins, S. (1992) TDM, AQD's, AVR's, SLO's and TMA's: An Introduction to Transportation Planning in Southern California. Paper presented at the Universities Transport Studies Group Conference, Newcastle upon Tyne.

Banister, D. (1977) Car Availability and Usage: A Modal Split Model based on these Concepts. University of Reading, Geographical Paper No 58.

Banister, D. (1978) The influence of habit formation on modal choice – a heuristic model. *Transportation*, **7**(1), pp. 5–18.

Banister, D. (1981) The response of the Shire Counties to the question of transport needs. *Traffic Engineering and Control*, **23**(10), pp. 488–491.

Banister, D. (1983) Transport needs in rural areas – A review and proposal. *Transport Reviews*, **3**(1), pp. 35–49.

Banister, D. (1985) Deregulating the bus industry in Britain – the proposals. *Transport Reviews*, **5**(2), pp. 99–103.

Banister, D. (ed.) (1989) The Final Gridlock. *Built Environment*, **15**(3/4), pp. 159–256.

Banister, D. (1990) Privatisation in transport: From the company state to the contract state, in Simmie, J. and King, R. (eds.) *The State in Action: Public Policy and Politics*, London: Pinter, pp. 95–116.

Banister, D. (1992*a*) Energy use, transport and settlement patterns, in Breheny, M. (ed.) *Sustainable Development and Urban Form*. London: Pion, pp. 160–181.

Banister, D. (1992*b*) The British Experience of Bus Deregulation in Urban Transport: Lessons for Europe. University College London, Planning and Development Research Centre, Working Paper 5.

Banister, D. (1992*c*) Demographic Structure and Social Behaviour. Proceedings of the 12th International Symposium on Theory and Practice in Transport Economics – Transport Growth in Question. Lisbon, pp. 109-150.

Banister, D. (1993) Policy responses in the UK, in Banister, D. and Button, K. (eds.) *Transport, the Environment and Sustainable Development*. London: Spon, pp. 53–78.

Banister, D. J., Bould, M. and Warren, G. (1984) Towards needs based transport planning. *Traffic Engineering and Control*, **25**(7/8), pp. 372–375.

Banister, D. and Botham, R. (1985) Joint land use and transport planning: The case of Merseyside, in Harrison, A. and Gretton, J. (eds.) *Transport UK 1985: An Economic, Social and Policy Audit*. Newbury: Policy Journals, pp. 95–100.

Banister, D. and Norton, F. (1988) The role of the voluntary sector in the provision of rural services – the case of transport. *Journal of Rural Studies*, **4**(1), pp. 57–71.

Banister, D. and Mackett, R. (1990) The minibus: Theory and experience, and their implications. *Transport Reviews*, **10**(3), pp. 189–214.

Banister, D. and Pickup, L. (1990) Bus transport in the Metropolitan Areas and London, in Bell, P. and Cloke, P. (eds.) *Deregulation and Transport: Market Forces in the Modern World*. London: Fulton, pp. 67–83.

Banister, D. and Bayliss, D. (1991) Structural Changes in Population and Impact on Passenger Transport Demand. European Conference of Ministers of Transport, Round Table 88. Paris, pp. 103–142.

Banister, D., Cullen, I. and Mackett, R. (1991) The impacts of land use on travel demand, in Rickard, J. and Larkinson, J. (eds.) *Longer Term Issues in Transport*. Aldershot: Avebury, pp. 81–130.

Banister, D., Berechman, J. and De Rus, G. (1992) Competitive regimes within the European bus industry: Theory and practice. *Transportation Research*, **26A**(2), pp. 167–178.

Batey, P. W. and Breheny, M. J. (1978) Methods in strategic planning. *Town Planning Review*, **49**, pp. 259–273, and pp. 502–518.

Batty, M. (1976) *Urban Modelling: Algorithms, Calibrations and Predictions*. Cambridge: Cambridge University Press.

Batty, M. (1981) A perspective on urban systems analysis, in Banister, D. and Hall, P. (eds.) *Transport and Public Policy Planning*. London: Mansell, pp. 421–440.

Batty, M. (1989) Urban modelling and planning: Reflections retrodictions and prescriptions, in Macmillan, B. (ed.) *Remodelling Geography*. Oxford: Basil Blackwell, pp. 147–169.

Beauregard, R. (1990) Bringing the city back in. *Journal of the American Planning Association*, **56**(2), pp. 210–215.

Beesley, M. and Kain, J. F. (1964) Urban form, car ownership and public policy: An appraisal of Traffic in Towns. *Urban Studies*, **1**(2), pp. 174–203.

Beesley, M. and Kain, J. F. (1965) Forecasting car ownership and use. *Urban Studies*, **2**(2), pp. 163–185.

Bendixson, T. (1989) Transport in the Nineties: The Shaping of Europe. A Study commissioned by the Royal Institution of Chartered Surveyors, London.

Benwell, M. (1980) The Contribution of the Social Sciences to Transport Research in France. Transport and Road Research Laboratory, SR 637.

Berlinski, D. (1976) *On Systems Analysis: An Essay Concerning the Limitations of Mathematical Methods in the Social, Political and Biological Sciences.* Cambridge, Mass: MIT Press.

Bernard, M. J. (1981) Problems in predicting market response to new transportation technology, in Stopher, P. R., Meyburg, A. H. and Brög, W. (eds.) *New Horizons in Travel Behaviour Research.* New York: Lexington Books, pp. 465–87.

Blowers, A. (1986) Town planning – paradoxes and prospects. *The Planner*, April, pp. 82–96.

Bluestone, B. and Harrison, B. (1982) *The Deindustrialization of America: Plant Closures, Community Abandonment, and the Dismantling of Basic Industry.* New York: Basic Books.

Boddy, M. and Thrift, N. (1990) Socio economic restructuring and changes in patterns of long distance commuting in the M4 corridor. Final report to the ESRC.

Bonnafous, A. (1987) The regional impact of the TGV. *Transportation*, **14**(2), pp. 127–137.

Boyce, D., Day, N. and McDonald, C. (1970) Metropolitan Plan Making, Monograph Series 4. Philadelphia, Regional Science Research Institute.

Breheny, M. (1989) The planning paradox: Strategic issues and pragmatic responses in the South of England. *Built Environment* **14**(3/4), pp. 225–239.

Breheny, M. (1991) The renaissance of strategic planning? *Environment and Planning B*, **18**(3), pp 233–249.

Breheny, M. and Roberts, A. J. (1981) Forecasts in structure plans. *The Planner*. July/August, pp 102–103.

Brewer, G. D. (1973) *Politicians, Bureaucrats and the Consultant: A Critique of Urban Problem Solving.* New York: Basic Books.

British Road Federation (1972) *Basic Road Statistics.* London: British Road Federation.

Brög, W. (1992) Structural Changes in Population and Impact on Passenger Transport – Germany. ECMT Round Table 88. Paris, pp. 5-42.

Brotchie, J. F., Dickey, J. W. and Sharpe, R. (1980) *TOPAZ – General Planning Technique and Its Application at Regional, Urban and Facility Planning Levels.* Berlin: Springer-Verlag.

Bruton, M. J. (1985) *Introduction to Transportation Planning.* London: UCL Press – Reprinted in 1992.

Burnett, P. and Hanson, S. (1982) The analysis of travel as an example of complex human behaviour in spatially-constrined situations: Definitions and measurement issues. *Transportation Research,* **16A**(2), pp 87–102.

Butler, E. W., Chapin, F. S., Hemmens, G., Kaiser, E. J., Stegman, M. A. and Weiss, S. F. (1969) Moving behavior and residential choice – A national survey. Highway Research Board, Special Report 81.

Button, K. (1991) Transport and communications, in Rickard, J. and Larkinson, J. (eds.) *Longer Terms Issues in Transport.* Aldershot: Avebury, pp 323–360.

California Transportation Commission (1987) *California Congestion: Its Effect Now and in the Future.* Sacramento, CA: California Transportation Commission.

Carlstein, T., Parkes, D. and Thrift, N. (eds.) (1978) *Timing Space and Spacing Time,* Volume 2, Human Activity and Time Geography. London: Edward Arnold.

Castells, M. (1983) *The City and the Grassroots: A Cross-Cultural Theory of Urban Social Movements.* London: Edward Arnold.

Castells, M. (1990) *The Informational City: Information, Technology, Economic Restructuring and the Urban Regional Process.* Oxford: Blackwell.

Cervero, R. (1989) *America's Suburban Centers: The Land-Use Transportation Link*. Boston: Unwin Hyman.

Cervero, R. and Hall, P. (1990) Containing traffic congestion in America. *Built Environment*, **15**(3/4), pp. 176–184.

Chen, K. (1990) Driver information systems: A North American perspective. *Transportation*, **17**(3), pp. 251–62.

Champion, A.G., Green, A.E., Owen, D. W., Ellis, D. J. and Coombes, M. G. (1987) *Changing Places: Britain's Demographic, Economic and Social Complexion*. London: Edward Arnold.

Chicago Area Transportation Study (1960) *Final Report*. Chicago.

Civil Aviation Authority (1989) *Traffic Distribution Policy for Airports Serving the London Area*. CAP 559. London: Civil Aviation Authority.

Coburn, T. M., Beesley, M. E. and Reynolds, D. J. (1960) The London-Birmingham Motorway: Traffic and Economics. Road Research Laboratory, Department of Scientific and Industrial Research, Technical Paper 46. London: HMSO.

Coindet, J. – P. (1988) Financing Urban Public Transport in France. Paper presented at the PTRC Annual Conference, University of Sussex.

Commission of the European Communities (CEC) (1985) *Official Journal of the European Communities*, C 144, 13 June.

Commission of the European Communities (CEC) (1990) *Green Paper on The Urban Environment*. EUR 12902. Brussels.

Commission of the European Communities (CEC) (1991) *The DRIVE Programme in 1991*. DGXIII DRIVE1202. Brussels.

Commission of the European Communities (CEC) (1992*a*) *Green Paper on The Impact of Transport on the Environment: A Community Strategy for Sustainable Development*. DGVII. Brussels.

Commission of the European Communities (CEC) (1992*b*) *The Future Development of the Common Transport Policy*. Com (92) 494. Brussels.

Coombe, R. D., Forshaw, I. G., Bamford, T. J. G. (1990) Assessment in the London Assessment Studies. *Traffic Engineering and Control*,**31**(10), pp. 510–518.

Coopers and Lybrand (1990) *Internationale vergelijking infrastructuur*. Rotterdam: Coopers and Lybrand.

Coopers and Lybrand, Deloitte (1990) *The Environmental White Paper: This Common Inheritance, Briefing Digest*. London: Coopers and Lybrand, Deloitte.

Creighton, R. L. (1970) *Urban Transportation Planning*. Chicago: University of Illinois Press.

Cullingworth, J. B. (1988) *Town and Country Planning in Britain*. London: Unwin Hyman.

Dahl, R. (1972) *Der Anfang von Ende des Autos* (The Beginning of the End of the Car). Munden: Langeswiesche Brandt.

Daly, A. (1981) Behavioural travel modelling: Some European experience, in Banister, D. and Hall, P. (eds.) *Transport and Public Policy Planning*, London: Mansell, pp. 307–316.

Davidoff, P. (1965) Advocacy and pluralism in planning. *Journal of the American Planning Association*, **31**(4), pp. 331–338.

Deakin, E. (1990*a*) Land use and transportation planning in response to congestion problems: A review and critique. *Transportation Research Record*, No. 1237, pp. 77–86.

Deakin, E. (1990*b*) Toll roads: A new direction for US highways? *Built Environment*, **15**(3/4), pp. 185–194.

Department of Energy (1990) *Energy Use and Energy Efficiency in UK Transport up to the Year 2010*. Energy Efficiency Office, Energy Efficiency Series 10. London: HMSO.

Department of the Environment (1972*a*) *Getting the Best Roads for Our Money: The COBA Method of Appraisal*. London: HMSO.

Department of the Environment (1972*b*) *New Roads in Towns*. London: HMSO.

Department of the Environment (1973) *Greater London Development Plan: Report of the Panel of Inquiry*. London: HMSO.

Department of the Environment (1975) *Transport Supplementary Grant Submission for 1977/78*. Circular 125/75. London: HMSO.

Department of the Environment (1983) *Streamling the Cities: Government Proposals for Reorganising Local Government in Greater London and the Metropolitan Counties*. Cmnd 9063. London: HMSO.

Department of the Environment (1986) *The Future of Development Plans: A Consultation Paper*. London: HMSO.

Department of the Environment (1989) *The Future of Development Plans*. Cm 569. London: HMSO.

Department of the Environment (1990) *This Common Inheritance: Britain's Environmental Strategy*. Cm 1200. London: HMSO.

Department of the Environment (1992) *Climate Change: Our National Programme for $CO_2$ Emissions*. A Discussion Document. London: Department of the Environment.

Department of Trade (1974) *Maplin: Review of Airport Project*. London: HMSO.

Department of Transport (1976*a*) *Transport Policy: A Consultation Document*, Vols I and II. London: HMSO.

Department of Transport (1976*b*) *Report on the Location of Major Inter-Urban Road Schemes with Regard to Noise and Other Environmental Issues* (Chairman: J. Jefferson). London: HMSO.

Department of Transport (1977*a*) *Transport Policy*: White Paper, Cmnd 6836. London: HMSO.

Department of Transport (1977*b*) *Report of the Advisory Committee on Trunk Road Assessment* (Chairman: Sir George Leitch). London: HMSO.

Department of Transport (1978*a*) *Policy for Roads: England 1978*. Cmnd 7132. London: HMSO.

Department of Transport (1978*b*) *Report on the Review of Highway Inquiry Procedures*. Cmnd 7133. London: HMSO.

Department of Transport (1979*a*) *Trunk Road Proposals – A Comprehensive Framework for Appraisal* (Chairman: Sir George Leitch). London: HMSO.

Department of Transport (1979*b*) *Road Haulage Operators' Licensing*. Report of the Independent Committee of Inquiry (Chairman: Sir C. Foster). London: HMSO.

Department of Transport (1980) *Report of the Inquiry into Lorries, People and the Environment* (Chairman: Sir A. Armitage). London: HMSO.

Department of Transport (1983) *Manual of Environmental Appraisal*. London: HMSO.

Department of Transport (1984) *Buses*. White Paper, Cmnd 9300. London: HMSO.

Department of Transport (1986*a*) *Urban Road Appraisal, Report of the Standing Advisory Committee on Trunk Road Appraisal*. (Chairman: Professor T. E. H. Williams). London: HMSO.

Department of Transport (1986*b*) *The Government Response to the SACTRA Report on Urban Road Appraisal*. London: HMSO.

Department of Transport (1988) *National Travel Survey 1985/86*. London: HMSO.

Department of Transport (1989*a*) *National Road Traffic Forecasts (Great Britain)*. London: HMSO.

Department of Transport (1989*b*) *Roads for Prosperity*. Cm 693. London: HMSO.

Department of Transport (1989c) *New Roads by New Means*. Cm 698. London: HMSO.
Department of Transport (1990) *Transport Statistics Great Britain 1979–1989*. London: HMSO.
Department of Transport (1991a) *Report 1991: The Government's Expenditure Plans 1991–92 to 1993–94*. Cm 1507. London: HMSO.
Department of Transport (1991b) *Bus and Coach Statistics: Great Britain, 1990/91*. London: HMSO.
Department of Transport (1992a) *Transport Statistics Great Britain 1992*. London: HMSO.
Department of Transport (1992b) *Assessing the Environmental Impact of Road Schemes*. Report of the Standing Committee on Trunk Road Appraisal. London: HMSO.
Dix, M. (1981) Contributions of psychology to developments in travel demand modelling, in Banister, D. and Hall, P. (eds.) *Transport and Public Policy Planning*. London: Mansell, pp. 369–386.
Dollinger, H. (1972) *Die Totale Autogesellschaft* (The Total Car Society). Passau: Carl Hanser.
Domencich, T. and McFadden, D. (1975) *Urban Travel Demand: A Behavioural Analysis*. Amsterdam: North Holland.
Duncan, S. S. (1990) Do housing prices rise that much? A dissenting view. *Housing Studies*, 5(3), pp. 195–208.
Dunleavy, P. and Duncan, K. (1989) Understanding the Politics of Transport. Paper presented at an ESRC Seminar on Policy Making Processes in Transport, University College London.
Dunleavy, P. and O'Leary, B. (1987) *Theories of the State: The Politics of Liberal Democracy*. London: Macmillan.
Dupuit, J. (1844) On the measurement of the utility of public work. *Annales des Ponts et Chaussées*, 2nd Series, Vol. 8. (Translation: Barback, R. H. (1952) *International Economic Papers*, 2, pp. 83–110.
Dupuy, G. (1978) *Une technique de planification au service de l'automobile*. Paris: Centre à la Recherche d'Urbanisme.
EC DRIVE II BATT Consortium (1992) Initial ideas on defining objectives. Internal Working Note 1. University College London.
EC Council Directive (1985) On the assessment of the effects of certain public and private projects on the environment. *Official Journal of the European Communities*, L175, pp. 40–48.
Echenique, M. H., Flowerdew, A. D. J., Hunt, J. D., Mayo, T. R., Skidmore, I. J. and Simmonds, D. C. (1990) The MEPLAN models for Bilbao, Leeds and Dortmund. *Transport Reviews*, 10(4), pp. 309–322.
Economist Intelligence Unit (1964) Urban transport planning in the United Kingdom. *Motor Business*, 40, pp. 17–32.
Ellickson, R. C. (1977) Suburban growth controls: An economic and legal analysis. *The Yale Law Journal*, 86(2), pp. 385–511.
Else, P. and Trinder, E. (1991) Transport priorities in local government. *Traffic Engineering and Control*, 32(5), pp. 261–263.
Engwight, D. (1992) *Towards an Eco-City: Calming the Traffic*. Sydney: Envirobook.
Environment and Transport Planning (1991) *Traffic Calming Manual*. Unpublished Consultants Report. Brighton: ETP.
Etzioni, A. (1967) Mixed scanning: A third approach to decision making. *Public Administration Review*, Winter, pp. 217–242.
European Conference of Ministers of Transport (1982) Review of Demand Models. Report of the ECMT Round Table 58. Economic Research Centre, Paris.
European Conference of Ministers of Transport (1992) High Speed Rail. Report of the

ECMT Round Table 87. Economic Research Centre, Paris.

European Investment Bank (1991) *Briefing Series: Communications*. Luxembourg: EIB.

European Parliament (1991) Community Policy on Transport Infrastructures. European Parliament, Research and Development Papers Series on Regional Policy and Transport 16. Luxembourg.

European Round Table of Industrialists (1988) *Need for Renewing Transport Infrastructure in Europe. Proposals for Improving the Decision Making Process*. Brussels: ERTI.

European Round Table of Industrialists (1990) *Missing Networks in Europe. Proposals for the Renewal of Europe's Infrastructure*. Brussels: ERTI.

Evans, S. H. and Mackinder, I. H. (1980) Predictive Accuracy of British Transport Studies. Paper presented at the PTRC Annual Conference. University of Warwick.

Fainstein, S. S. (1991) Promoting economic development: Urban planning in the United States and Great Britain. *Journal of the American Planning Association*, **57**(1), pp. 22–33.

Faludi, A. (1987) *A Decision Centred View of Environmental Planning*. Oxford: Pergamon.

Federal Highway Administration (1988) America's Challenge for Highway Transportation in the 21st Century. Interim Report on the Future National Highway Program Taskforce.

Federal Highway Authority (FHWA) (1984) *Highway Statistics*. Washington DC: US Department of Transportation.

Ferreira, L. J. (1981) Long Term Urban Transportation Planning Methodology in France: A Review. Institute for Transport Studies, University of Leeds, WP 151.

Foglesong, R. E. (1986) *Planning the Capitalist City*. Princeton, NJ: Princeton University Press.

Foot, D. (1981) *Operational Urban Models: An Introduction*. London: Methuen.

Forester, J. (1980) Critical theory and planning practice. *Journal of the American Planning Association*, **46**(2), pp. 275–286.

Forrester, J. W. (1969) *Urban Dynamics – City Growth, Stagnation and Decay*. Cambridge, Mass: MIT Press.

Foster, C. D. (1974) *Lessons of Maplin: Is the Machinery of Government Decision-Making at Fault?* London: Institute of Economic Affairs.

Foster, C. D. and Beesley, M. E. (1963) Estimating the social benefit of constructing an underground railway in London. *Journal of the Royal Statistical Society A*, **126**(1), pp. 46–92.

Frank Graham Consulting Engineers (1991) *Royal Borough of Kensington and Chelsea Traffic Calming Study*. Hertford: Frank Graham Consulting Engineers.

Freeman Fox, Wilbur Smith and Associates (1968) *West Midlands Transport Study*. Report for Local Authorities, Ministry of Transport, Ministry of Housing, and Local Transport Organizations.

Friedmann, J. (1973) Retracking America: A Theory of Transactive Planning. Garden City: Doubleday.

Gakenheimer, R. (1976) *Transportation Planning as a Response to Controversy: The Boston Case*. Cambridge, Mass: MIT Press.

Gent, H. A. and Nijkamp, P. (1989) Devolution of Transport Policy in Europe. Report produced as part of the ESF-NECTAR activities. Amsterdam.

Gerardin, B. (1989a) Possibilities for, and Costs of, Private and Public Investment in Transport. European Conference of Ministers of Transport, Round Table 80, Paris.

Gerardin, B. (1989b) Financing the European High Speed Rail Network. Paper presented at the PTRC Annual Conference, University of Sussex.

Gerardin, B. and Viegas, J. (1992) European Transport Infrastructure and Networks: Current

Policies and Trends. Paper presented at the NECTAR International Symposium, Amsterdam.

Girnau, G. (1983) Wo Kann Gespart Werden im U- und Stadtbahnbau. *Der Nahverkehr*, 1/83, pp 8–16.

Glaister, S. (1990) Bus Deregulation in the UK. Draft paper presented to the World Bank, p. 36.

Goddard, J. (1989) Urban development in an information economy: Policy issues. OECD Group on Urban Affairs, Project on Urban Impacts of Technological and Socio-Demographic Change, UP/TSDC (89) 1. Paris: OECD.

Goldner, W (1971) The Lowry model heritage. *Journal of the American Institute of Planners*, **37**(1), pp. 100–110.

Goodwin, P. (1990) Demographic impacts, social consequences, and the transport policy debate. *Oxford Review of Economic Policy*, **6**(2), pp. 76–90.

Goodwin, P. (1991) The right tools for the job: A research agenda, in Rickard, J. and Larkinson, J. (eds.) *Longer Term Issues in Transport*. Aldershot: Avebury, pp. 1–40.

Goodwin, P. and Layzell, A. (1985) Longitudinal analysis for public transport policy issues, in Jansen, G. R. M., Nijkamp, P. and Ruijgrok, C. J. (eds.) *Transportation and Mobility in an Era of Transition*. Amsterdam: North Holland, Elsevier, pp. 185–200.

Goodwin, P., Hallett, S., Kenny, F. and Stokes, G. (1991) Transport: The New Realism. Report to the Rees Jeffreys Road Fund. Oxford.

Gordon, P. and Richardson, H. W. (1989) Gasoline consumption and cities: A reply. *Journal of the American Planning Association*, **55**(3), pp. 342–346.

Government Statistical Office (1990) *Social Trends 1990*. London: HMSO.

Grandjean, A. and Henry, C. (1984) Economic rationality in the development of a motorway network. *Transport Reviews*, **4**(2), pp. 143–158.

Greene D. L. (1987) Long run vehicle travel prediction from demographic trends. *Transportation Research Record*, 1135, pp. 1–9.

Grieco, M., Pickup, L. and Whipp, R. (eds.) (1989) *Gender, Transport and Employment: The Impact of Travel Constraints*. Aldershot: Avebury.

Group Transport 2000 Plus (1990) Transport in a Fast Changing Europe. Report produced by the Group set up by Karel Van Miert, Transport Commissioner of the European Commission.

Gwilliam, K. M. (1990) A European perspective: Long term research needs in the Netherlands, in Rickard, J. and Larkinson, J. (eds.) *Longer Term Issues in Transport*. Aldershot: Avebury, pp. 245–304.

Hagerstrand, T. (1970) What about people in regional science? *Papers of the Regional Science Association*, **24**(1), pp. 7–21.

Hall, P. (1980) *Great Planning Disasters*. London: Weidenfeld and Nicholson.

Hall, P. (1985) Urban transportation: Paradoxes for the 1980s, in Jansen G. R. M., Nijkamp, P. and Ruijgrok, C. J. (eds.) *Transportation and Mobility in an Era of Transition*. Amsterdam: Elsevier, North Holland, pp. 367–375.

Hall, P. (1988) *Cities of Tomorrow: An Intellectual History of Urban Planning and Design in the Twentieth Century*. Oxford: Blackwell.

Hall, P. (1989) The turbulent eighth decade: challenger to American city planning. *Journal of the American Planning Association*, **55**(2), pp. 275–282.

Hall, P. (1991) Three systems, three separate paths. *Journal of the American Planning Association*. **57**(1), pp. 16–20.

Hall, P. and Hass-Klau, C. (1985) *Can Rail Save The City?* Farnborough: Gower.

Hall, P. and Markusan, A. (1985) *Silicon Landscapes*. London: George Allen and Unwin.

Hamnett, C., Harmer, M. and Williams, P. (1991) *Safe as Houses: Housing Inheritance in Britain*. London: Paul Chapman Publishing.

Hanson, M. E. (1992) Automobile subsidies and land use: Estimates and policy responses. *Journal of the American Planning Association*, **58**(1), pp. 60–71.

Harvey, D. (1978) On planning the ideology of planning, in Burchall, R. W. and Sternlieb, G. (eds.) *Planning Theory in the 1980s*. New Brunswick, NJ: Centre for Policy Research, Rutgers University.

Hass-Klau, C. (1990) *The Pedestrian and City Traffic*. London: Belhaven.

Hay, A. (1992) Equity in transport planning? *The Planner*, **78**(7), pp. 12–13.

Hay, A. and Trinder, E. (1991) Concepts of equity, fairness and justice expressed by local transport policy makers. *Environment and Planning C*, **9**(4), pp. 453–465.

Heggie, I. (1978) Putting behaviour into behavioural models of travel choice. *Journal of the Operational Research Society*, **29**(6), pp. 541–550.

Hemmens, G. C. (1968) Survey of planning agency experience with urban development models, data processing and computers. SR97. Highway Research Board, Washington DC.

Hensher, D. A. (1989) Behavioural and resource values of travel time savings: A bicentennial update. *Australian Road Research*, **19**(3), pp. 223–229.

Hepworth, M. and Ducatel, K. (1992) *Transport in the Information Age: Wheels and Wires*. London: Belhaven.

Hills, M. (1973) Planning for Multiple Objectives. Monograph No 5. Regional Science Research Institute, Philadelphia PA.

HM Treasury (1989) *Government Plans*. Cm 621. London: HMSO.

Hollatz, J, W. and Tamms, F. (eds.) (1965)*Die Kommunalen Verkehrsproblem in der Bundesrepublik Deutschland* (The Traffic Problems of Local Authorities in the Federal Republic of Germany). Essen: Vulkan.

Holliday, I., Marcou, G. and Vickerman, R. (1991) *The Channel Tunnel: Public Policy, Regional Development and European Integration*. London: Belhaven.

Hoos, I. R. (1972) *Systems Analysis in Public Policy: A Critique*. California: University of California Press.

House of Commons (1990) *Roads for the Future*. Report of the Transport Committee. Session 1989–1990, HC 198, Vols I and II. London: HMSO

House of Commons Expenditure Committee (1973) *Urban Transport Planning*, Vols I,II,III. London: HMSO

House of Lords Select Committee on Science and Technology (1987) *Innovation in Surface Transport*. HL 57-I. London: HMSO.

Hutchinson, B. (1974) *Principles of Urban Transport Systems Planning*. New York: McGraw Hill.

Hutchinson, B. (1981) Urban transport policy and policy analysis methods. *Transport Reviews*, **1**(2), pp. 169–188.

Institute of Civil Engineers (1983) *Report of the Civil Engineering Task Force on Public Inquiry Procedures* (Chairman: Lord Vaizey). London: Institute of Civil Engineers.

Institute of Transportation Engineers (1980) Evaluation of the accuracy of past urban transportation forecasts. *ITE Journal*, February, pp. 24–34.

Jones, D., May, A. and Wenban-Smith, A. (1990) Integrated transport studies: Lessons from the Birmingham Study. *Traffic Engineering and Control*, **31**(11), pp. 572–576.

Jones, P. (1990) (ed.) *New Developments in Dynamic and Activity Based Approaches to Travel Analysis*. Aldershot: Gower.

Jones, P., Clarke, M. and Dix, M. (1983) *Understanding Travel Behaviour*. Aldershot: Gower.

Kain, J. F. (1978) The use of computer simulation models for policy analysis. *Urban Analysis*, **5**(2), pp. 175–189.

Kashima, S. (1989) Advanced traffic information systems in Tokyo. *Built Environment*, **15**(3/4), pp. 244–250.

Kawashima, H. (1990) Japanese perspective of driver information systems. *Transportation*, **17**(3), pp. 263–84.

Kay, J. A. and Thompson, D. (1986) *Privatisation and Regulation – The UK Experience*. Oxford: Oxford University Press.

Keating, G. (1992) Toll tales on the highway to prosperity. *Independent*, 16 December.

Khattak, A. J., Schofer, J. L. and Koppelman, F. S. (1991) Commuters' Enroute Diversion and Return Decisions: IVHS Design Implications. Proceedings of the 6th International Conference on Travel Behaviour, Quebec, pp 362–76.

Kitamura, R. (1990) Panel analysis in transportation planning: an overview. *Transportation Research*, **24A**(6), pp. 401–415.

Kostyniuk, L. P. and Kitamura, R. (1987) Effects of aging and motorization on travel behavior: An Exploration. *Transportation Research Record*, 1135, pp. 31–36.

Kroes, E. P. and Sheldon, R. J. (1988) Stated preference methods. An introduction. *Journal of Transport Economics and Policy*, **22**(1), pp. 11–26.

Kutter, E. (1972) *Demographische Determinanten Stadtlischen Personenverkehrs*. Braunschweig.

Lancaster, K. J. (1966) A new approach to consumer theory. *Journal of Political Economy*, **84**(2), pp. 132–157.

Laslett, P. (1990) *A Fresh Map of Life: The Emergence of the Third Age*. London: Weidenfeld and Nicholson.

Lassave, P. and Offner, J. M. (1989) Urban transport: changes in expertise in France in the 1970s and 1980s. *Transport Reviews*, **9**(2), pp. 119–135.

Lave, C. (1990) Things won't get a lot worse: The future of US traffic congestion. University of California, Irvine, Economics Department, Mimeo.

Lawless, P. (1986) *The Evolution of Spatial Policy*. London: Pion.

Lee, D. (1973) Requiem for large scale models. *American Institute of Planners*, **39**(2), pp. 163–178.

Leibbrand, K. (1964) *Verkehr und Stadtebau*. Basel: Birkhauser Verlag.

Lenntorp, B. (1981) A time-geographic approach to transport and public policy planning, in Banister, D. and Hall, P. (eds.) *Transport and Public Policy Planning*. London: Mansell, pp. 387–396.

Levin, P. H. (1979) Highway inquiries: A study in government responsiveness. *Public Administration*, **57**(1), pp. 21–49.

Lex Motoring (1992) Lex Report on Motoring 1992. Report produced by MORI for Lex Motoring, London.

Lichfield, N. (1970) Evaluation methodology of urban and regional plans: A review. *Regional Studies*, **4**(2), pp. 151–165.

Lichfield N. (1992) Planning and markets in transport, in Glaister, S., Lichfield N., Bayliss, D., Travers, T. and Ridley, T.(eds.) *Transport Options for London*. London: London School of Economics, Greater London Group, pp. 142–160.

Lichfield, N., Kettle, P. and Whitbread, M. (1976) *Evaluation in the Planning Process*. Oxford: Pergamon Press.

Lindblom, C. E. (1959) The science of muddling through. *Public Administration Review*, Spring, pp. 151–169.

Lojkine, J. (1974) *La politique urbaine dans la Region Lyonnaise 1945–72*. Paris: Mouton.

London County Council (1964) *London Traffic Survey*. London: LCC.

London Strategic Policy Unit (1986) Transport Policies for London 1987–88. Transport Policy Group.

Louvière, J. (1988) Conjoint analysis modelling of stated preferences. A review of theory, methods, recent developments and external validity. *Journal of Transport Economics and Policy*, **22**(1), pp. 93–120.

Lowry, I. (1964) *A Model of Metropolis*, RM-4035-RC. Santa Monica, California: Rand Corporation.

Lundqvist, L. (1984) Analysing the impacts of energy factors on urban form, in Brotchie, J. F., Newton, P. W., Hall, P. and Nijkamp, P. (eds.) *The Future of Urban Form*. London: Croom Helm.

Mäcke, P. (1964) *Das Prognoseverfahren in der Strassenverkehrsplanung*. Berlin: Bau Verlag.

Mackett, R. L. (1985) Integrated land use-transport models, *Transport Reviews*, **5**(4), pp. 325–343.

Mackett, R. L. (1990) Comparative analysis of modelling land-use transport interaction at the micro and macro levels. *Environment and Planning A*, **22**, pp. 457–475.

McLoughlin, J. B. (1969) *Urban and Regional Planning: A Systems Approach*. London: Faber and Faber.

Maggi, R., Masser, I. and Nijkamp, P. (1992) Missing networks in European transport and communications. *Transport Reviews*, **12**(4), pp. 311–321.

Markusan, A. (1985) *Profits, Cycles, Oligopoly and Regional Development*. Cambridge, Mass: MIT Press.

Masser, I., Sviden, O. and Wegener, M. (1992) *The Geography of Europe's Futures*. London: Belhaven.

May, A. (1991) Integrated transport studies: A new approach to urban transport policy formulation in the UK. *Transport Reviews*, **11**(3), pp. 223–248.

Meyer, J. R., Kain, J. F. and Wohl, M. (1965) *The Urban Transportation Problem*. Cambridge, Mass: Harvard University Press.

Michelman, F. I. (1967) Property, utility and fairness: Comments on the ethical foundation of 'Just Compensation' law. *Harvard Law Review*, **80**(6), pp. 1165–1258.

Miles, I. (1989) The electronic cottage: Myth or reality? *Futures*, **21**(1), pp. 47–59.

Ministry of Transport (1963) *Traffic in Towns: A Study of the Long Term Problems of Traffic in Urban Areas* (Chairman: Colin Buchanan). London: HMSO.

Ministry of Transport (1964) *Road Pricing: The Economic and Technical Possibilities* (Chairman: R. Smeed). London: HMSO.

Ministry of Transport (1967) *Better Use of Town Roads*. London: HMSO.

Ministry of Transport (1968) *Traffic and Transport Plans*. Roads Circular 1/68. London: HMSO.

Ministry of Transport (1969) *Roads for the Future: A New Inter-Urban Plan*, London: HMSO.

Ministry of Transport (1970) *Roads for the Future: The New Inter-Urban Plan for England*. Cmnd 4369. London: HMSO.

Mitchell, R. and Rapkin, C. (1954) *Urban Traffic—A Function of Land Use*. New York: Columbia University Press.

Monheim, R. (1980) *Fussgangerbereiche und Fussgangerverkehr in Stadtzentren*. Bonn: Bonner Geographische Abhandlungen.

Moss, M. L. (1987) Telecommunications, world cities and urban policy. *Urban Studies*, **24**(4), pp 534–541.

MVA (with ITS Leeds and TSU Oxford) (1987) *The Value of Time Savings*. Newbury: Policy Journals.

National Audit Office (1988) *Road Planning*. House of Commons Paper 688. London: HMSO.

National Committee on Urban Transportation (1958) *Better Transportation for Your City: A Guide to the Factual Development of Urban Transportation Plans*. Chicago: Public Administration Service.

National Economic Development Council (NEDC) (1991) Amber Alert: Relieving

Urban Traffic Congestion, Report of the Traffic Management Systems Working Party, Chaired by John Ashworth.

National Economic Development Office (NEDO) (1984) *A Fairer and Faster Route to Major Road Construction*. London: NEDO.

Newman, P. W. G. and Kenworthy, J. R. (1989) *Cities and Automobile Dependence*. Aldershot: Gower.

Nilles, J. M. (1988) Traffic reduction by telecommunications: A status review and selected bibliography. *Transportation Research*, **22A**(4), pp. 301–317.

Noortman, H. (1988) The changing context of transport and infrastructure policy. *Environment and Planning C*, **6**(2), pp. 131–144.

Orfeuil, J.-P. (1991) Structural Changes in Population and Impact on Passenger Transport Demand. Paper presented at the European Conference of Ministers of Transport, Round Table 88. Paris.

Organisation for Economic Cooperation and Development (OECD) (1989) Comparative socio demographic trends: Initial data on ageing populations. OECD Group on Urban Affairs, UP/TSDA(89)1. Paris: OECD.

Organisation for Economic Cooperation and Development (OECD) (1991) *Environmental Indicators*. Paris: OECD.

Pack, J. R. (1978) Urban Models: Diffusion and Policy Applications, Monograph Series 7, Philadelphia, Regional Science Research Institute.

Pangalos, S. (1989) Prominent Applications and Software. Paper presented at the DRIVE-EUROFRET Workshop on Freight and Fleet Management. Brussels.

Paradeise, C. (1980) The Contribution of Social Sciences to Transport Research in Germany. Institute de Recherche des Transports, Paris.

Pendyala, R. M., Goulias, K. G. and Kitamura, R. (1991) Impact of Telecommunications on Spatial and Temporal Patterns of Household Travel: An Assessment for the State of California Pilot Projects Participants. Prepared for the California State Department of Transportation, UCD-ITS-RR-91-07.

Plowden, S. (1972) *Towns Against Traffic*. London: Andre Deutsch.

Plowden, W (1973) *The Motor Car and Politics in Britain*, Harmondsworth: Penguin.

Polak, J. (1987) A comment on Supernak's critique of transport modelling. *Transportation*, **14**(1), pp. 63–72.

Prest, A. R. and Turvey, R. (1965) Cost benefit analysis: A survey. *Economic Journal*, **75**, pp. 685–705.

Prevedouros, P. D. and Schofer, J. L. (1989) Suburban transport behaviour as a factor in congestion. *Transportation Research Record*, 1237, pp. 47–58.

Priemus, H. and Nijkamp, P. (1992) Randstad policy on infrastructure and transportation: High ambitions, poor results, in Dieleman, E. M. and Musterd, S. (eds.) *The Randstad: A Research and Policy Laboratory*. Amsterdam: Kluwer Academic, pp. 165–191.

Public Accounts Committee (1989) *Road Planning*. 15th Report of the PAC, HC 101. London: HMSO.

Putman, S. H. (1983) *Integrated Urban Models*. London: Pion.

Putman, S. H. (1986) Integrated Policy Analysis of Metropolitan Transportation and Location. Report DOT-P-30-80-32. US Department of Transportation, Washington DC.

Reade, E. (1987) *British Town and Country Planning*. Milton Keynes: Open University Press.

Rees, R. (1986) Is there an economic case for privatisation? *Public Money*, **5**(4), pp. 19–26.

Reward Group (1990) Annual Survey of Employee Benefits. Mimeo. London.

Richmond, J. E. D. (1990) Introducing philosophical theories to urban transport planning. *Systems Research*, **7**(1), pp. 47–56.

Rickaby, P. (1987) Six settlement patterns compared. *Environment and Planning B*, **14**(3), pp. 193–223.

Rietveld, P. (1993) Policy responses in the Netherlands, in Banister, D. and Button, K. (eds.) *Transport, the Environment and Sustainable Development*. London: Spon, pp. 102–113.

Roos, D. and Altshuler, A. (1984) *The Future of the Automobile*. London: George Allen and Unwin.

Rosenbloom, S. and Altshuler, A. (1977) Equity issues in urban transportation. *Policy Studies Journal*, **6**(1), pp. 29–39.

Roskill, Lord Justice (1971) *Report of the Commission on the Third London Airport*. London: HMSO.

Savitch, H. (1988) *Post Industrial Cities*. Princeton, NJ: Princeton University Press.

Sayer, R. A. (1976) A critique of urban modelling. *Progress in Planning*, **6**(3), pp. 187–254.

Schon, D. A. (1983) *The Reflective Practitioner*. New York: Basic Books.

Scott, M. (1969) *American City Planning Since 1890: A History Commemorating the Fiftieth Anniversary of the American Institute of Planners*. Berkeley: University of California Press.

Self, P. (1975) *Econocrats and the Policy Process: The Politics and Philosophy of Cost-Benefit Analysis*. London: Macmillan.

Self, P. (1980) *Planning the Urban Region*. London: George Allen and Unwin.

Simpson, B. (1987) *Planning and Public Transport in Great Britain, France and West Germany*. London: Longman.

Smith, P. (1988) *City, State and Market*. Oxford: Blackwell.

Starkie, D. N. M. (1973) Transportation planning and public policy. *Progress in Planning*, **1**(4), pp. 313–389.

Starkie, D. N. M. (1982) *The Motorway Age: Road and Traffic Policies in Post-War Britain*. Oxford: Pergamon.

Stigler, G. J. (1975) *The Citizen and the State: Essays on Regulation*. Chicago: University of Chicago Press.

Supernak, J. (1983) Transportation modelling: lessons from the past and tasks for the future. *Transportation*, **12**(1), pp. 79–90.

Supernak, J. and Stevens, W. R. (1987) Urban transportation modelling: the discussion continues. *Transportation*, **14**(1), pp. 73–82.

Tanaka, M. (1991) Dealing with unanticipated events. *The Wheel Extended*, **76**, p 30.

Taylor, M. D. P., Young, W., Wigan, M. R. and Ogden, K. W. (1992) Designing a large scale travel demand survey: New challenges and new opportunities. *Transportation Research*, **26A**(3) pp. 247–261.

Tennyson, R. (1991) Interactions between Changing Urban Patterns and Health. Paper presented at the Solar Energy Conference on Architecture in Climate Change, RIBA.

Thomson, J. M. (1969) *Motorways in London*. London: Duckworths.

Thomson, J. M. (1974) *Modern Transport Economics*, Harmondsworth: Penguin.

Toffler, A. (1991) *Power Shift: Knowledge, Wealth and Violence at the Edge of the 21st Century*. London: Bantam.

Tolley, R. (1990) *Calming Traffic in Residential Areas*. Tregaron, Wales: Brefi Press.

Traffic Research Corporation (1969) *Merseyside Area Land-use Transportation Study*. Final Report Vol A.

Transnet (1990) *Energy, Transport and the Environment*. London: Transnet.

Trinder, E., Hay, A., Dignan, J., Else, P. and Skorupski, J. (1991) Concepts of equity, fairness and justice in British Transport legislation 1960–1988. *Environment and Planning C*, **9**(1), pp. 31–50.

Tyson, W. J. (1988) A Review of the First Year of Bus Deregulation. Report to the Association of Metropolitan Authorities and Passenger Transport Executive Group. Manchester.

Tyson, W. J. (1989) A Review of the Second Year of Bus Deregulation. Report to the Association of Metropolitan Authorities and Passenger Transport Executive Group. Manchester.

Tyson, W. J. (1990) Effects of deregulation on service coordination in the Metropolitan Areas. *Journal of Transport Economics and Policy*, **24**(3), pp. 283–294.

Tyson, W. J. (1991) Urban public transport fare structures, in Rickard, J. and Larkinson, J. (eds.) *Longer Term Issues in Transport*. Aldershot: Avebury, pp. 305–322.

UK Treasury (1984) *Investment Appraisal in the Public Sector: A Technical Guide for Government Departments*. London: HM Treasury.

US Department of Transportation Federal Highways Administration (1968) Instructional Memorandum 50-4-68: Operations Plans for 'Continuing' Urban Transportation Planning. Washington DC.

US Department of Transportation (1986) Personal travel in the US, in *The Nationwide Personal Transportation Study*, Vol 2. Washington DC.

Van Rest, D. (1987) Policies for major roads in urban areas in the UK. *Cities*, **4**(3), pp. 236–252.

Vibe, N. (1991) Are There Limits to Growth in Car Ownership? Some Recent Figures from the Oslo Region. Preliminary paper presented at the ECMT Round Table 88. TP/0371/1991.

Vickerman, R. W. (1991) Transport infrastructure in the European Community: New developments, regional implications and evaluation, in Vickerman, R. W. (ed.) *Infrastructure and Regional Development*. London: Pion, pp. 36–50.

Vogel, S. and Rowlands, I. (1990) The challenges and opportunities facing the European electronics information industry, in Locksley G. (ed.) *The Single European Market and the Information and Communication Technologies*. London: Belhaven.

Wachs, M. (1985a) The politicization of transit subsidy policy in America, in Jansen, G. R. M., Nijkamp, P. and Ruijgrok, C. J. (eds.) *Transportation and Mobility in an Era of Transition*. Amsterdam: Elsevier, North Holland, pp. 353–366.

Wachs, M. (ed.) (1985b) *Ethics in Planning*. New Brunswick, NJ: Centre for Urban Policy Research, Rutgers University.

Wachs, M. (1990) Regulating traffic by controling land uses: The Southern California experience. *Transportation*, **16**(3), pp. 241–256.

Webber, M. M. (1976) The BART experience – What have we learned? Monograph 26. Institute of Urban and Regional Development and Institute of Transportation Studies, University of California, Berkeley.

Webster, M. (1984) *Explanation, Prediction and Planning: The Lowry Model*. Pion: London.

Webster, F. V. and Paulley, N. J. (1990) An international study on land use and transport interaction. *Transport Reviews*, **10**(4), pp. 287–308 and *Transport Reviews*, **11**(3), pp. 197–222.

Weiner, E. (1985) Urban transportation planning in the US: An historical overview. *Transport Reviews*, **4**(4), pp. 331–358 and **5**(1), pp. 19–48.

Wermuth, M., Kutter, E., Zumkeller, D., Heidemann, C. and Brog, W. (1990) Federal Republic of Germany, in Nijkamp, P., Reichman, S. and Wegener, M. (eds.) *Euromobile: Transport, Communications and Mobility in Europe*. Farnborough: Avebury, pp. 185–196.

Whitt, J. A. (1982) *Urban Elites and Mass Transportation: The Dialectics of Power*. Princeton, NJ: Princeton University Press.

Wood, C. and Jones, C. (1991) *Monitoring Environmental Assessment and Planning.* Report by the EIA Centre, Department of Planning and Landscape, University of Manchester, for the Department of the Environment. London: HMSO.

Wood, D. (1992) Environmental quality and value for money, in Banister, D. and Button, K. (eds.) *Transport, the Environment and Sustainable Development.* London: Spon, pp. 212–216.

Wrigley, N. (1986) Quantitative methods: the era of longitudinal data analysis. *Advances in Human Geography*, **10**, pp. 84–102.

Yiftachel, O. (1989) Towards a new typology of urban planning theories. *Environment and Planning C*, **16**(1), pp. 23–39.

Zopel, C. (1991) A New Transport Policy. Friedrich Ebert Foundation, Economic and Social Policy in Europe, Background Paper No 1. Berlin.

# Index